Science and the Life-World

SCIENCE AND THE LIFE-WORLD

Essays on Husserl's 'Crisis of European Sciences'

Edited by David Hyder and
Hans-Jörg Rheinberger

Stanford
University
Press

Stanford,
California

Stanford University Press
Stanford, California

Printed in the United States of America on acid-free, archival-quality paper

Library of Congress Cataloging-in-Publication Data

Science and the life-world : essays on Husserl's Crisis of European sciences / edited by David Hyder and Hans-Jörg Rheinberger.
p. cm.
Includes bibliographical references and index.
ISBN 978-0-8047-5604-4 (cloth : alk. paper)
1. Husserl, Edmund, 1859–1938. Krisis der europäischen Wissenschaften und die transzendentale Phänomenologie. 2. Husserl, Edmund, 1859–1938—Influence. 3. Science—Philosophy. 4. History—Philosophy. 5. Phenomenology. I. Hyder, David Jalal, 1954– II. Rheinberger, Hans-Jörg.
B3279.H93K74 2010
142'.7—dc22 2009029815

Typeset by Westchester Book Group in 10.9/13 Adobe Garamond

Contents

Contributors

David Carr (Ph.D. Yale, 1966) has taught at Yale, the University of Ottawa (Canada), and, since 1991, at Emory University, where he was department chair and is now Charles Howard Candler Professor. He is the author of *Phenomenology and the Problem of History* (1974), *Time, Narrative and History* (1986), *Interpreting Husserl* (1987), and *The Paradox of Subjectivity* (1999).

Eva-Maria Engelen is currently working in an interdisciplinary research project on consciousness at the Berlin-Brandenburgische Akademie der Wissenschaften in Berlin and teaching at the University of Konstanz, Germany. She has several publications in the history of philosophy as well as in the philosophy of language, epistemology, and the theory of emotions. Her major publications are *Zeit, Zahl und Bild. Studien zur Verbindung von Philosophie und Wissenschaft bei Abbo von Fleury* (1993), *Das Feststehende bestimmt das Mögliche* (1999), *Erkenntnis und Liebe* (2003), *Descartes* (2005) and *Gefühle* (2007).

Dagfinn Føllesdal, C.I. Lewis Professor of Philosophy, Stanford University, studied science and mathematics in Oslo and Göttingen (1950–57), philosophy at Harvard (Ph.D., Harvard, 1961), then taught at Harvard and later at Oslo (1967–99) and Stanford (1966–present). His publications include *Husserl und Frege* (1958), *Referential Opacity and Modal Logic* (2004), and other books and articles in philosophy of language, philosophy of science, and on contemporary continental philosophy. He was editor of *The Journal of Symbolic Logic* from 1970 to 1982.

Michael Friedman is currently Frederick P. Rehmus Family Professor of Humanities at Stanford University. His publications include *Foundations of Space-Time Theories: Relativistic Physics and Philosophy of Science* (1983), *Kant and the Exact Sciences* (1992), *Reconsidering Logical Positivism* (1999), *A Parting of the Ways: Carnap, Cassirer, and Heidegger* (2000), and *Dynamics of Reason* (2001).

Rodolphe Gasché is Eugenio Donato Chair of Comparative Literature and SUNY Distinguished Professor at the State University of New York at Buffalo. He studied philosophy and comparative literature in Munich, Berlin, and Paris. He holds an M.A. and Ph.D. in philosophy from the Freie Universität Berlin (Germany). His interests concern nineteenth- and twentieth-century French literature, critical theory, and its relation to continental philosophy since early romanticism. His publications include *Die hybride Wissenschaft* (1973), *The Tain of the Mirror: Derrida and the Philosophy of Reflection* (1986), *Inventions of Difference: On Jacques Derrida* (1994), *The Idea of Form: Rethinking Kant's Aesthetics* (2003), and *Europe, or The Infinite Task: A Study of a Philosophical Concept* (2009).

Ian Hacking is professor emeritus at the Collège de France and University of Toronto. His books include, on probability ideas, *The Emergence of Probability* (1975) and *The Taming of Chance* (1990); on mental disorders, *Rewriting the Soul* (1995) and *Mad Travelers* (1998); on questions of general interest, *Why Does Language Matter to Philosophy?* (1975) and *The Social Construction of What?* (1999); and on experimental science, *Representing and Intervening* (1983). His collection of papers, *Historical Ontology* (2002), suggests the types of philosophical topics in which he is interested.

Michael Hampe is professor of philosophy at the Swiss Federal Institute of Technology Zurich. He works on the history of early modern philosophy (Hobbes, Hume and Spinoza), the history of twentieth-century metaphysics (Whitehead, Dewey, and Sellars) and on problems of the philosophy of biology and psychology. His latest books are *Erkenntnis und Praxis. Zur Philosophie des Pragmatismus* (2006), and *Eine kleine Geschichte des Naturgesetzbegriffs* (2007).

David Hyder is associate professor of philosophy at the University of Ottawa. He studied philosophy at Yale, Toronto, and Göttingen, was a postdoctoral researcher at the Max Planck Institute for the History of Science, and was later assistant professor of philosophy at the University of Konstanz, where he did his habilitation on Kant and Helmholtz. He is the author of *The Mechanics of Meaning* (2002) and *The Determinate World* (2009).

Ulrich Majer is professor at the Philosophisches Seminar of the University of Göttingen. He is general editor of *David Hilbert's Lectures on the Foundations of Mathematics and Physics, 1891–1933* of which have appeared *David Hilbert's Lectures on the Foundations of Geometry, 1891–1902* (Springer 2004), and *Relativity, Quantum Theory and Epistemology, 1915–1927* (Springer 2009).

Hans-Jörg Rheinberger studied philosophy and biology in Tübingen and Berlin. Since 1997, he has been director at the Max Planck Institute for the History of Science in Berlin. He has published numerous articles on molecular biology and the history of science, in particular the history and epistemology of experimentation. Among his books are *Toward a History of Epistemic Things: Synthesizing Proteins in the Test Tube* (1997), *Iterationen* (2005), *Epistemologie des Konkreten* (2006), and *Historische Epistemologie zur Einführung* (2007).

David Woodruff Smith is associate professor at the University of California. He specializes in phenomenology and ontology, with an eye to philosophy of mind and language and history of twentieth-century philosophy. His books include *Husserl* (2007), *Mind World* (2004), *The Circle of Acquaintance* (1989), and *Husserl and Intentionality* (1982, coauthored with Ronald McIntyre). He is the editor of *Phenomenology and Philosophy of Mind* (2005, coedited with Amie L. Thomasson) and *The Cambridge Companion to Husserl* (1995, coedited with Barry Smith).

Friedrich Steinle is professor of history of science and technology and director at the Interdisciplinary Center of Science and Technology Studies: Normative and Historical Perspectives at the University of Wuppertal, Germany. His research interests concern the history and philosophy of science, with a special focus on experiment, and case studies of seventeenth- to nineteenth-century physical science. His books include *Newtons Manuskript 'De gravitatione'* (1991) and *Explorative Experimente: Ampère, Faraday und die Ursprünge der Elektrodynamik* (2005). He is coeditor of *Revisiting Discovery and Justification: Historical and Philosophical Perspectives on the Context Distinction* (2006, with J. Schickore).

Acknowledgments

The editors would like to thank Antje Radeck and Josephine Fenger for their organizational help, Leona Geisler for her invaluable assistance in preparing the manuscript, and Jean-Luc Greenwood and Lucas Jurkovic for their handling of the bibliographical research. They would also like to thank Cathy Brown for her superb copyediting and Cynthia and Robert Swanson for the index. They are grateful to the criticisms and suggestions of an anonymous referee, as well as for support from the University of Ottawa and the Max Planck Institute for the History of Science in Berlin.

Introduction

David Hyder

The essays composing this volume are all concerned with Edmund Husserl's last work, *The Crisis of the European Sciences and Transcendental Phenomenology*, a book that was meant to cap Husserl's long and illustrious career, and yet was never published in its entirety in his lifetime. To date, little attention has been paid to the *Crisis*, which is remarkable when one considers Husserl's reputation during the first half of the twentieth century. Under normal circumstances, its publication would have been greeted with great interest. In fact, few works in the history of philosophy have had such a dismal reception, which begins in neglect, in Nazi Germany, and turns to rejection in the 1950s and 1960s, when both continental and analytic philosophers were opposed to many of its most basic assumptions. What is the source of this aversion? And why should we devote our time to the book now? These introductory remarks provide brief answers to these questions and are followed by a discussion of the individual contributions.

The book was born in a political crisis and, as Eva-Maria Engelen explains in her contribution to this volume, at a time of personal crisis as well. Husserl was invited to give lectures in 1935 in both Vienna and Prague, titled "Philosophy in the Crisis of European Mankind" and "The Crisis of European Sciences and Psychology," respectively. These lectures were later reworked as the first two sections of the *Crisis* and were published in the journal *Philosophia* in 1936, shortly before Husserl became ill. They were supplemented by the third section when Walter Biemel finally published them in 1954 in their final, but still fragmentary, form. The original publication was almost totally ignored, although the publication in France in 1939 of what later became Appendix VI, *The Origin of Geometry*, did provoke some discussion there.

Husserl used the occasion of the Vienna and Prague lectures to address topics that had come to preoccupy him in the increasingly hostile environment in which he found himself. His state of mind can be inferred from comments made to Gustav Albrecht in the following year, in which he makes clear that he has been driven to take a stand on his own status as a *German* philosopher. "I have finally at least to make clear for myself," writes Husserl, "that I am no stranger in German philosophy . . . , and that all of the great thinkers of the past would have had to count me as a true heir of their spirit, as blood of their blood."[1] His defense against ostracism is to insist first on the primacy of the European philosophical tradition and second on his belonging to that tradition. He responds to opponents who would banish him from Europe by championing it. Philosophical culture is European culture, and European culture corresponds quite exactly to what we refer to as *the West*. We learn that "spiritual" Europe includes some of the "colonies of England" but only those parts that are predominantly white. There are apparently no contributions to the history of science and philosophy from the Arabs, Indians, or Chinese. This is not a popular line to take today, whatever truth there might be to it.

In retrospect, we can see that this political goal was doomed from the start, and thus it is tempting to read the work today without regard to it. Since Husserl was *forced* to make a claim that we find obvious—that he was a German philosopher, heir to the tradition of German Idealism—it can scarcely be enlightening to watch him defend that truism against the absurdities of his National Socialist opponents. It is tragic, or at the very least embarrassing, to find Husserl employing the language of *Blut und Boden* when defending himself against the Nazis. And we are easily induced to adopt the same attitude toward the *arguments* he uses. He should not have to stoop so low.

It is worth emphasizing how much the origins of this text in Nazi Germany affect its content, because these aspects may elicit negative impressions today. They contributed to its peculiar status in the postwar period, when it emerged as a foil to what I call, in my contribution to this volume, the French linguistic turn. In this later context, reaction to the *Crisis* was less political and more philosophical, but it remained critical. And since developments in France were part and parcel of a large international movement characterized by suspicion of mental or cognitive explanations, there was little interest or sympathy shown for Husserl's text in the English-speaking world of the 1950s and 1960s outside the embattled and increasingly isolated group of American phenomenologists. As phenomenology petered out in the analytic world, as behaviorism, Quine, and Wittgenstein carried the day, there was no place left for this peculiar phenomenological coda. Lastly, it did not help that, among phenomenologists, Husserl's *Crisis* came to be seen by some as a last-ditch attempt on Husserl's part to keep up with his protégé Heidegger, whose existential phenomenology was breaking new ground. With the possible exception of the school of Constructivism centered at the University of Erlangen, it is safe to say that by the 1970s, this was a book almost wholly unloved (a notable exception is Elisabeth Ströker, 1979, 1997).

But we should be reading it now. The wheel has turned, and many of the doctrines that postwar philosophers rejected, and on whose ground they rejected Husserl, have returned in force. There are systematic reasons for this. On the analytic side, the sustained arguments of Searle and others have forced the theory of intentionality back on the agenda, and there is increasing recognition that Husserl provides us with the most developed modern theory of intentionality. We now realize that the gap between continental and analytic philosophy has limited applicability before the 1940s: As David Woodruff Smith observes, Husserl and Carnap are correctly viewed as sharing problems and methods, for instance the notion of a constitution-system. Above all, the problem lying at the heart of the *Crisis*, namely the relation between the world as experienced and the world as it exists scientifically, is now the subject of lively debate. Recognition of intentionality as a legitimate subject for research has meant readmitting the referents of everyday intentions (think of qualia), that is to say readmitting something very much like Husserl's life-world. One need only consider "theory theory" and the naive science program, both of which seek to analyze the belief systems underlying speakers' everyday reasoning about the world and other people.

External factors also impinge on the philosophical debate. The progress of the sciences has accelerated; the gap we experience between the life-world and the world of science is wider. At the same time, the scientific world intrudes in the life-world ever more obviously, further weakening the latter's claim to legitimacy. Finally, even those aspects of the book that may seem politically incorrect to contemporary readers are topical and merit serious reflection. As Michael Friedman, David Carr, and Rodolphe Gasché all point out in their contributions to this volume, Husserl does not simply champion the European tradition. He subjects it to the same critical inquiry that it has always required of itself, and he detects a paradox in the fact that this supposedly universal form of thought is nevertheless specific to a given cultural and historical setting, and that it may therefore be unjustified in its claim to universal validity.

Like the writings of the late Wittgenstein, the *Crisis* is difficult to read because it presumes working knowledge of the complex development in Husserl's philosophical thinking that preceded it. And it is not just a recapitulation; it introduces new interpretative problems of its own. Because it is in many respects the summation of Husserl's project, it will hardly come as a surprise that this is a book about everything: about the world and the self, though perhaps that is normal for theoretical philosophy, but also about a large part of the history of philosophy and science. In this regard it is nothing short of Hegelian, even if its author was suspicious of Hegel. The essays collected in this volume attempt to provide some of the context necessary for understanding this difficult work. The breadth of topics addressed in the *Crisis* is reflected in the scope of the papers, which are divided into three main areas: Husserl's theories of science and the life-world, the theory of history advanced in the *Crisis*, and the dissemination and application of Husserl's views.

Husserl's *Crisis* is not a work in the philosophy of science as we now understand the term; however, it is deeply concerned with science and its development, with its relation to everyday experience, and with its internal structure. In this sense, the *Crisis* can be correctly said to have a theory of science. In his "Science, Intentionality, and Historical Background," David Woodruff Smith extends ideas drawn from Husserl to consider how scientific theories are grounded in intentional activities occurring in concrete historical contexts, including both the natural world they represent and the human world in which they are developed. On Smith's view, the early Husserl in *Logical Investigations* took a theory to be a form of "pure logic," while in the *Crisis*, the late Husserl worried how our idealized theory of nature–mathematical physics–had become alienated from our commonsense understanding of the world.

According to Smith, a scientific theory is a structure of propositions supported by evidence. These propositions are intentional contents of scientific thought or judgment, formed from concepts developed in the course of scientific research and tied through perceptual evidence to concepts developed in everyday life. Though ideal meanings, theoretical propositions are bound to historical circumstances in several ways. In his chapter, Smith appraises these types of dependence on historical context, addressing both the indexical character of scientific concepts or propositions and the historical background of theorizing they presuppose.

Smith shows how an extended Husserlian view of science elucidates (a) Husserl's seemingly divergent early and late views of theory, (b) Quine's "web" view of scientific theory (with mathematics and logic at the center of the web of belief), and (c) Michael Friedman's revisionary Kantian-Carnapian view of relativized a priori principles within a scientific theory. The basic, formal ontology of a scientific theory is, for that range of theory, a priori (Friedman), the least revisable part of the theory (Quine), and furthest from our ordinary range of *Lebenswelt* (life-world) theory (Husserl). Yet on Smith's view, a theory remains a system of ideal propositions bound by dependence to the world in which we put forth these propositions.

One benefit of Smith's analysis consists in identifying the central role of the life-world in Husserl's theory—a concept some have seen as new to the *Crisis*. In his "The *Lebenswelt* in Husserl," Dagfinn Føllesdal analyzes what he sees as the long role of this key Husserlian notion in phenomenology, with particular attention to the notions of noema, horizon, constitution, and world. On Føllesdal's view, the notion of a life-world derives from a distinction between the "natural" and "transcendental" attitudes that Husserl introduced in 1906, and which he develops under a variety of names, until it makes its first appearance in 1917. This life-world is pregiven and intersubjective, for it contains other people with their own intentional states and attitudes toward the same world we all share. Føllesdal maintains that these essential attributes of the life-world remain unchanged in the *Crisis*, and that its appearance there conforms to Husserl's past understanding. Husserl saw it as opening a new way to approach phenomenology, in that the latter can be understood in part as an attempt to characterize the formation or construction of this natural world.

Føllesdal believes we should resist the urge to see the life-world as something set up *against* science, for the scientific world is itself part of the life-world, scientific statements get their meaning from being embedded in the life-world, and thus they are also ultimately *justified* by the life-world. He concludes by considering this last key function of the concept, which appears to hold for Husserl even in the case of mathematics. How can the life-world be the source of ultimate justification if it is also the realm of *doxa*, of mere opinion and prejudice? Føllesdal's reply is that for Husserl, there is no dishonor in so justifying our beliefs, because there is no alternative. Even the scientist must revert in ultimate justification to the everyday world and life he lives.

This interpretation finds agreement in Ulrich Majer's discussion of the views of Hermann Weyl, who claims that the scientist or mathematician always has recourse to primitive life-world capacities. Majer's "The Origin and Significance of Husserl's Notion of the *Lebenswelt*" approaches the life-world from the point of view of contemporaneous philosophy of science, above all the philosophical writings of Weyl and David Hilbert, two prominent Göttingen mathematicians and physicists indirectly connected with Husserl. He argues that Husserl's concept is an important *discovery*—both philosophically as well as from a scientific point of view. For Majer, it would be a mistake to read the *Crisis* as a work on the history of philosophy, enriched with some history of science; rather, we are dealing with philosophy dressed up as history. He identifies the key philosophical problem as akin to what Husserl's contemporary Weyl had described as a "ridiculous circle." To illustrate the circle, Weyl imagines a reductive sequence in which we analyze "the chalk on the blackboard" as an amalgam of elementary particles, and these in turn are "dissolved" into a rule-governed calculus, which is in turn revealed to be nothing other than "concrete signs written with chalk on the blackboard." Weyl contends that this shows us that there is an *irreducible* basis to our experience, which he refers to as "the manner in which we understand man and things in daily life" (Weyl, 1968, 342).

It is in fact one of the central themes of the *Crisis* that we cannot understand science with the methods of science on pain of circularity. But there is a deeper connection to be seen, Majer suggests, when one recalls that Weyl is thinking here in part of the foundations of mathematics, in particular Hilbert's formalism. In such a pure formalism, Weyl argues, we can dispense with almost all familiar (including logical) concepts, but we still presume basic capacities to operate with symbols. It would be wrong, he asserts, "to reduce this naively and roughly understood spatial order of signs to a *purified* spatial intuition and structure as it is expressed for example in Euclidean geometry." This capacity is a "natural understanding in dealing with things in our natural world surrounding us" (1968, 341). To hope to replace our natural understanding of signs with a theoretical account would be, Majer suggests, a category mistake akin to the confusion of symbol and sign, which distinction is itself a reflection of the distinction between mind and body.

This close agreement between Husserl's and Weyl's views raises the question of whether anyone has actually ever fallen into Weyl's ridiculous circle. Many scientists, Majer points out, are sensitive to these issues. But there are also scientists who believe that all traces of the mental can be banished from our scientific ontology, a position that is of course familiar to many philosophers as strong reductionism or physicalism. To champion this strong view, Majer selects David Hilbert, the scientist who maintained that all problems that can be clearly stated can be solved. If Hilbert is aware of the circle, that would be good reason to believe that most scientists are. And here we may be reassured, for Hilbert himself acknowledges that there is something paradoxical to the idea of a complete scientific theory, one that explains everything, for such a theory would transform our thought itself into something "merely apparent."

The foundations of mathematics are also the main concern of Ian Hacking's "Husserl on the Origins of Geometry," which focuses on *The Origin of Geometry*, an appendix to the *Crisis* that has received almost as much attention as the work itself. While Hacking is sympathetic to Husserl, he is above all critical of the latter's notion of sedimentation, which on Hacking's view is far less common in mathematics than Husserl seems to think.[2] According to Hacking, Husserl is one of those philosophers who is awed by mathematics, particularly by mathematical proof, which is not surprising when one recalls that his doctorate was in mathematics, not philosophy. This awe is bound up with Husserl's historico-foundational project, however; that is to say with the search for what Hacking calls the *Ur* words: the *Ursprung* (primal origins) of geometry, its *Urstiftung* (primal establishment). The search for origins betrays a tendency found also in Freud, ultimately due to Hegel, namely to think that by finding the origin of some event or process (pathology, proof), we will somehow grasp or undo the problems it presents to us. This is a project that Hacking himself cannot share, because, in his view, the ability to make proofs is an innate human capacity, which makes a search for the unique historical origin of mathematical proof otiose.

Hacking develops these ideas by contrasting Husserl's views on the origins of geometry with Kant's, drawing on the well-known passage from the preface to the first *Critique*, in which Kant imagines the first geometer. Unlike Kant's lone geometer, Husserl's form a community; and whereas Kant focuses on the notion of conceptual construction (mathematical proof), Husserl directs his attention to the formation of ideal objects. His primitive geometers construct an ideal ontology through their communal linguistic activity. Lastly, whereas Kant's scientific hero is Newton, Husserl focuses on Galileo, or more precisely on *his* Galileo, what he sometimes calls the "Galilean style." Hacking disagrees with Husserl, however, who he thinks unjustly saddles Galileo with an old ontology of ideal geometrical objects, while what Galileo mathematized were not *things* but *motions*.

Hacking concludes by considering Husserl's "historical a priori" as well as the process of sedimentation. With regard to the former, he takes a "French" line, seeing in Husserl the source of Foucault's homonymous concept, namely historical structures of thought or speech that determine "conditions of possibility and exclusion that come into being" in a given historical frame.[3] This notion is clearly related to that of sedimentation, in that we call such a priori structures historical precisely because they are now concealed from us. To reactivate them is to unearth sedimented norms. It is on just this point that Hacking deviates most strongly from Husserl's views, for he simply does not accept that mathematical knowledge has the sedimentary structure that Husserl supposes. Considering a number of contemporary examples, he argues that modern mathematicians often have "the living experience of the very evidence, the primal evidence, that Husserl sought." This is because Kant had it better than Husserl: Geometric and algorithmic reasoning are indeed a cultural phenomenon, according to Hacking, but this is because they rest on innate human capacities, which may express themselves whenever the circumstances are favorable, and which therefore do not have, as Husserl believed, an inaugural moment.

The next group of essays concern the second major theme I mentioned in introduction, namely the theory of history implicit in the *Crisis*. How is Husserl's preoccupation with history to be understood? Does it constitute a philosophy of history, and if so, in what sense? How does it compare with other approaches to history? In his contribution, David Carr develops answers to these questions by drawing on the customary distinction between substantive or speculative philosophy of history, on the one hand, and critical or analytical philosophy of history on the other. Although he allows that there are elements of each in the *Crisis*, he concludes that its most important contribution to the philosophy of history lies in the concept of *Geschichtlichkeit,* or historicity.

On the surface, Husserl's historical method is reminiscent of Hegel's, in that he offers a teleological history of Europe that is simultaneously a history of reason. But it would be a mistake, argues Carr, to read him in this way. The mere appearance of the term *crisis* ought be sufficient to warn us: Husserl views such a Hegelian project "as an object of bittersweet nostalgia and with a sense of loss," for the sciences are in crisis—they are no longer a source of hope and salvation for humankind in the way that Hegel envisaged. Nor will it do to read Husserl as substantive history in the mold of Kant, for here again there is no crisis that demands urgent intervention, whereas when we look to the philosophical literature in the decades preceding the *Crisis*, we find a number of similarly titled works with similar preoccupations. Carr concludes that Husserl's view of history is typical for the early twentieth century, even while the specific form of Husserl's crisis is intimately connected to his personal situation as a Jew (though a longtime convert) in Nazi Germany and to the situation of German philosophy as a whole at this time.

Carr next considers the analytical and epistemological aspects of Husserl's theory: Can it also be regarded as a contribution to the critical philosophy of history? The key distinction here is between the *Natur-* and *Geisteswissenschaften* (natural and human sciences) endorsed by German philosophers and historians in opposition to the historical positivism of Comte or Mill, which would assimilate history to natural science. Husserl sided with his countrymen on this issue, as reflected, for example, in his parallel distinction between the "naturalistic attitude" and the "personalistic attitude." But most of Husserl's subtle treatment of these issues is missing from the *Crisis*, according to Carr, and its contribution to the epistemology of the (human) sciences remains limited.

Husserl's real contribution to the philosophy of history is instead to be found in his notion of *Geschichtlichkeit,* or historicity, a term also prominent in Heidegger, and which may have been derived from Dilthey, according to whom we "are historical beings before we become observers of history" (1968, 277–78). For Husserl, this came to mean that the ego, as a self-constituting synthesis of temporal relations, constructs a narrative history of its own life; that is to say, it is an aspect of consciousness that it have historicity. But we do not obtain historicity proper until we have intersubjectivity, at which point the background of the individual's past becomes a social past, so that consciousness now includes history as an integral part.

Although historicity is not an epistemological concept, it does have an epistemological aspect, because it explains at the very least our *interest* in the past. Carr suggests that we can grasp the epistemological import of Husserl's notion if we compare it with what he says about the life-world. Just as the latter is the background without which there would be no natural science—a background both independent of and prior to the sciences—so the essential historicity of consciousness provides the background without which there would be no historical questions in the first place, whatever the specific epistemological problems these may raise.

Carr argues, however, that the epistemological implications of the concept go further. For involvement in any science entails an awareness of its historical and communal past, which is most immediately exemplified by Husserl's own awareness of the past history of philosophy, in which he as a philosopher will *intervene*, as opposed to merely describing or analyzing. This is, Carr suggests, a late realization on Husserl's part, for he had earlier insisted, for instance in *Ideas I*, that in doing phenomenology "we completely abstain from judgment respecting the doctrinal content of all pre-existing philosophy, and conduct all our investigations under this abstention" (§ 32). In his late work, Husserl comes to acknowledge that historical prejudices are not *mere* prejudices, that is to say that they too require the phenomenologist's attention. This fact leads to one of the central paradoxes of the *Crisis*, which concerns philosophy's history as a European tradition. For this tradition, as Husserl points out repeatedly, is a tradition that understands itself as free of any prejudices, thus presumably also those that derive from *its* specific historical situation. According to Carr, Husserl can no longer affirm the absolutely universal character of philosophy, but he is similarly aware that in conceiving of philosophy as a European tradition, he opens it to the charge of cultural relativism.

This deliberate tension, which remains unresolved in the *Crisis*, is a central topic of Michael Friedman's "Science, History, and Transcendental Subjectivity in Husserl's *Crisis*." Friedman traces out the long-standing connection between Galilean mathematical science and the phenomenological philosophy that begins with Husserl's 1910 "Philosophy as Rigorous Science" and extends to the *Crisis*. In that early text, Husserl took Galileo's work as a model for phenomenology: Just as mathematical physics requires an a priori mathematical structure in order to qualify as a rigorous science, so too philosophy must have an a priori part if it is to avoid the trap of naturalist psychology, and this a priori core is what will be provided by phenomenology. But this analogy between the two disciplines later becomes a problem, for by then the scientific ideal embodied in the Galilean style has caused a crisis at the heart of the sciences. The crisis results from our (that is, for Husserl, European humanity's) mistaking the ideal ontology of mathematical physics, which depends ultimately on life-world experiences, for the totality of objective reality. This mistake is only possible because of the sedimentation of geometric knowledge, that is to say the mechanical repetition of this knowledge without regard to its eidetic origins. Its most immediate consequence is Cartesian mind-body dualism, which cleaves the world into physical and objective matter and psychic and subjective mind.

The Cartesian split leaves the mind as a realm of appearances, and this conclusion can only be fought by adopting a position much like that advocated by Husserl in "Philosophy as Rigorous Science," namely one must recognize the transcendental character of mind, the fact that it has an a priori structure. But Husserl rejects Kant's attempts in this direction because they do not recognize the essentially perceptual or intuitive nature of mind: Kant's transcendental logic is arrived at not by considering the structure of experienced consciousness (even if Kant claims that he does), but by analyzing the sciences of logic and nature themselves and positing their rudiments as basic structures of cognition. Put simply, Kant ignores the life-world in his rush to ground Aristotelian logic and Newtonian physics.

From here, Friedman turns to an extended analysis of Husserl's notion of the life-world. He considers in particular the distinction between objective sciences following the Galilean style and the science of the life-world, which prescinds from all the special kinds of intuition appropriate to each of these sciences, dealing instead with the foundational structure that is the precondition of all of these—a process Husserl terms "the epoché of objective science." This *epoché*, or bracketing, is to be contrasted with the "Cartesian" method of the *Ideas*, in which we prescind from the entire "external" world, including the life-world itself. But it is not yet the full transcendental epoché Husserl develops in the *Crisis*, which requires changing our orientation within the life-world to reveal its transcendental structure, namely its intentional constitution.

It is here that Husserl's views are most clearly distinguished from Descartes' and Kant's. According to Friedman, Husserl is at pains in the *Crisis* to correct a common misapprehension of his project, namely that it is foundational. As Husserl himself points out, the aim of transcendental phenomenology "is not to secure objectivity but to understand it" (*Crisis*, § 55). This is not Descartes' foundationalist project; it is a transcendental one, which seeks to *ascertain* the conditions of experience. But it is not Kant's project either, because Husserl will not stop at what he calls "the phenomenological-psychological reduction," which is centered on the individual ego. The final transcendental reduction must acknowledge that "self-consciousness and consciousness of others are inseparable" (*Crisis*, § 71), by revealing that I am structured not as an individual consciousness, I am "within an inter-human present and within an open horizon of humankind: I know myself to be factually within a generative framework, in the unitary flow of an historical development" (ibid.).

Like Carr, Friedman argues that in arriving at this terminal position, Husserl also broke for good with the conception of transcendental phenomenology he had articulated on the Galilean model in "Philosophy as Rigorous Science." He does so because on that earlier view, the relation of phenomenology to psychology was analogous to that of mathematics to physics: Phenomenology would articulate the normative structure of psychology. But the new view includes intersubjectivity and historicity as essential elements of consciousness, and it is no longer concerned with the normative structure of an individual ego's thought alone. Instead, the role of transcendental phenomenology is now to intervene in the present crisis and recapture the original idea of science as a philosophical form of human existence.

This same paradox concerning philosophical and scientific universality is the central topic of Rodolphe Gasché's "Universality and Spatial Form." Gasché argues that Husserl's *Crisis* can be correctly understood as a critique of universality, more particularly of the failure of modern science to deliver on the promise of universal knowledge that was made in ancient Greece. This failure is also a failure of the European ideal, since

Europe, for Husserl, is inextricably linked to Western science. Universal science in the Greek sense is not only the science of the one true (as opposed to apparent) world, it is also a science freed from any particular attachments, whether these be to a given subject matter or to a given knower or community of knowers. But this attitude is effectively lost in the Renaissance, in that the *critical* aspect of this project—calling all particular attachments and traditions into doubt—is offset by an uncritical acceptance of its scientific fruits. Instead of being a truly universal form of knowledge, European science becomes anchored in the specifics of European culture so that the eventual "Europeanization of all other civilizations" becomes a "a historical non-sense" (*Crisis*, § 6). According to Gasché, what is exported as techno-science is a kind of universality that is (paradoxically) *specific* to European culture, thus one whose world is not truly shared by all.

The quintessential product of Greek universal science is geometry, the science of the ideal forms of space, which by determining the spatial forms of all possible (relative) worlds also determines the form of the one we all share. In the Renaissance, this universal aspect of geometry is used as a basis on which new universal sciences are constructed, thus paving the way for the mathematization of nature, a development that Husserl characterizes as "strange," because Greek geometric objects were *nonsensible idealities* (that was the source of their very universality) developed by abstracting from sensible, life-world experience. And yet for Galileo, this connection to the life-world is ignored in favor of geometry's universal, intersubjective aspects, resulting in "a tradition empty of meaning" (*Origin of Geometry*, p. 366). Gasché goes on to demonstrate that, for Husserl's Galileo, the privileged role of geometry derives from positing the one property all bodies have in common—shape—as fundamental, so that the remaining material properties of our experience (the so-called plena) *must be reducible in principle* to this one universal property. This results in the idealization of the whole of our experience, a goal Husserl calls "the Galilean idea," and thus in the loss of those very material properties that originally provided the basis for the universal idealities (*Crisis*, § 9e). According to this idea, the world is in principle completely mathematizable, and yet it remains a "strange" hypothesis, which is subject, in the course of natural science, to an endless process of verification. Modern scientific thought, instead of seeing this unending project as a means of framing the world, objectifies it—it takes "for *true being* what is actually a *method*" (*Crisis*, § 9h). This is, according to Gasché, "an ethico-philosophical error" that Husserl believes can only be corrected by recognizing the concealed project of universalization and idealization that began in ancient Greece, thereby acknowledging the essential historicity of geometric and natural scientific knowledge. For Husserl, that means acknowledging that they are the product of a constituting consciousness, whose study is precisely an object of phenomenological philosophy.

Eva-Maria Engelen investigates this same basic motif of the *Crisis*—the search for "original" experience or meaning, here exemplified in Greek scientific origins —in her "Husserl, History, and Consciousness." She considers three philosophically important origins of meaning in Husserl's text, and a fourth, personal one: consciousness, the life-world, European philosophy and science, and Husserl's personal origins as a Jewish and German thinker. The last two are evidently interconnected, since in writing the *Crisis*, Husserl is simultaneously establishing his identity as a *German* philosopher—against those who would deny him—by showing that he is an inheritor of the European philosophical tradition. In order to be such a descendant, he must himself engage in critical reflection on that tradition itself, since it has always been a tradition of critical inquiry. Engelen compares Husserl on this point to Michel Foucault, who argues that Kant was the first to pose the question concerning rationality as a *historical* question, that is, as one concerning the social and historical conditions of its development, as opposed to its origins in consciousness, considered atemporally. Husserl deliberately runs these two together: His search for the origins of European rationality is rooted both in present consciousness and past history. Engelen suggests that this historical/ahistorical duality is typical of all four of the origins of meaning that she considers. She concludes by considering the ways in which the concept of consciousness itself has a historical dimension, in that it did not and perhaps could not have existed for Greek thought. This raises the question of whether the Greeks could have experienced a life-world, or whether the latter is actually a post-Cartesian concept that has been illegitimately projected by Husserl into prehistory.

The duality between historical and ahistorical origins forms the basis of Michael Hampe's "Science, Philosophy, and the History of Knowledge" as well. Hampe considers Husserl's *Crisis*, and in particular the concept of a life-world, from the point of view of Wilfrid Sellars's later notions of the "manifest" and "scientific images." According to Hampe, both Husserl and Sellars were among the first to see something that has only recently begun to preoccupy philosophers generally, namely that a complete account of knowledge must make room for both objective and subjective knowledge and experience, as well as resolving the deep tension that always exists between the two. Both philosophers saw the resolution of this tension as lying unavoidably in philosophy, which thereby acquires a singular importance for human self-knowledge.

Hampe observes that both Sellars and Husserl follow in a line of philosophical argument that emerged in German idealism, in which the origins of human knowledge are described in terms that are superficially historical, but are in fact intended to be rational-dialectical. Like Fichte, Schelling, and Hegel, our two philosophers tell origin stories (the origin of man's concept of himself, the origin of geometry) that are clearly not intended to be actual history, but that instead explain the ideal essence of some unknown, and in their details, unimportant historical events. The fictitious *historical priority* of the life-world or the manifest image explains and justifies their *epistemological priority*, namely the claim that the scientific view is itself a product or development of them. This epistemological priority also shows itself in the present, in that the manifest properties of the life-world (for example, colors) persist as immediate data for the subject even once their natural scientific explanation has been given.

According to Hampe, both Husserl and Sellars suppose that in articulating the concealed elements of our knowledge—that is, the life-world background of the modern scientific view—we will achieve unification through transparency: By revealing hidden or "sedimented" knowledge-structures, we will supply the elements whose absence hinders unification. But in setting up the unity of knowledge as the goal to be achieved, they fail to recognize that much (scientific) knowledge is tacit and may therefore be in principle beyond "reactivation." Finally, it is just this project of unifying by revealing concealed knowledge-structures that necessitates the peculiar view of history just outlined: Each philosopher sees it as a development toward unification, as a process with a "hidden aim." Hampe has doubts similar to Engelen's about the assumptions bound up with Husserl's own conceptual tools. This whole approach would only make sense, Hampe argues, if historical investigation itself were completely transparent and did not rely on implicit knowledge of its own—and yet this is almost certainly not the case. Indeed, this assumption is reflective of a general tendency to underestimate the complexity and resilience of nonscientific knowledge and institutions, a tendency that might, on Hampe's view, "foster a kind of anthropological essentialism"—one that might, however, resemble that advocated by Hacking in his contribution when discussing mathematical proof.

The last three essays of this collection concern the application and dissemination of the late Husserl's theory of science. David Hyder discusses the relations between Husserl and later French authors such as Cavaillès and Foucault, who transform Husserl's transcendental project into a form of linguistic analysis. Friedrich Steinle shows how Husserl's ideas may be fruitfully extended and applied in historical research in the experimental sciences, in this case, chemistry. And in his "On the Historicity of Scientific Knowledge," Hans-Jörg Rheinberger considers Husserl's work as an instance of wider developments in the first decades of the twentieth century, in particular what has been called a "crisis of reality" or a "crisis of historicism." This double crisis, which Rheinberger argues is found in the work of Gaston Bachelard and Ludwik Fleck as well, led to a transformation in the history of science, whereby it became a properly epistemological enterprise and no longer merely the chronicler of positive scientific developments.

According to Rheinberger, late nineteenth-century thinkers, for example, Emil Du Bois-Reymond or Ludwig Boltzmann, held to a curious fusion of mechanism and historicism, in which historical explanation in the life sciences meshed with a naturalization of history, the link being Darwin's theory of evolution. But this view was only of brief duration, for two developments within the sciences themselves led to a new perspective on history, one exemplified in Husserl's and Fleck's writings. These developments were the rapid changes within contemporary physics and the same problem of the unity of the sciences just mentioned in our discussion of Hampe. The first of these led to an increasing awareness of the open-ended nature of scientific investigation, the fact that every theory is provisional. The second was given its bite both by the peculiar status of biological explanation, which employed concepts apparently distinct from physics, and by the internal differentiation within biology, both of which suggested that biology was not to be united with the rest of science easily, even if the Vienna Circle had made unification their watchword.

This is the context in which Ludwik Fleck, whose work is also discussed by Steinle, argued for an epistemology of the sciences that is both social and historical. Fleck's contemporary Bachelard drew similar lessons from developments in quantum theory: We should conceive of science as the activity of a community of human subjects who interact with the world in an ongoing process of investigation, as opposed to one in which they reflectively contemplate it. At that point, the sciences themselves become an object that can be investigated with historical methods. Rheinberger argues that we can understand Husserl's *Crisis* as motivated by shared questions concerning the (new) nature of the natural sciences, particularly the problem of accounting for human subjectivity in the terms of the natural sciences. We have seen this problem before: Natural science itself is a product of human activity, so this means explaining science by means of science, the very circularity Hermann Weyl called "ridiculous."

The absurdity of such circularity leads Husserl to ask for a "critique and clarification" of "that huge piece of method . . . that leads from the intuitively given surrounding world to the idealization of mathematics and to the interpretation of these idealizations as objective being" (*The Vienna Lecture*, p. 295). According to Rheinberger, Husserl understands himself to be supplying the necessary critique and clarification by means of his historico-epistemological investigation. But he could not free himself from foundational presuppositions, so that his historical investigations presuppose a teleological order to history and terminate in a foundational life-world. It is only in more recent work on the historical epistemology of the sciences that these foundational presuppositions are finally abandoned.

David Hyder's "Foucault, Cavaillès, and Husserl on the Historical Epistemology of the Sciences" deals with the impact of Husserl's *Crisis* on subsequent French philosophy of science, in particular the work of Jean Cavaillès and Michel Foucault. It is in France that the *Crisis*, and perhaps even more so, *The Origin of Geometry* had their greatest influence, in part due to the publication in 1939 of the latter text, with a series of commentaries, in a widely read special issue of the *Revue internationale de philosophie*. The substantial effect of this publication on the work of Jacques Derrida is well known;[4] however, its influence on the school of history and philosophy of science grouped around Gaston Bachelard is less well documented. Hyder shows that Jean Cavaillès drew heavily on

Husserl's account of scientific development in his *Sur la logique et la théorie des sciences*, a book that deeply affected Georges Canguilhem and his student, Michel Foucault. He argues that, in the hands of these French critics, the overall form of Husserl's account of sedimentation is preserved, but a fundamental thesis shared with Kant—that the history of science is in some sense a history of consciousness—is explicitly denied. The result is that while the sedimentation of scientific ontologies and their historical a prioris is an essential part of these French critics' approach, Foucault in particular insists that the structures in question are linguistic, as opposed to mental. He refers to the process of unearthing these hidden normative structures as "archaeology." Hyder argues that even the latter notion can be traced back to Eugen Fink's commentary on *The Origin of Geometry* in that same 1939 publication. The result is a Husserlian philosophy of science that has taken a linguistic turn remarkably similar to that which took place simultaneously in the English-speaking world.

Thus Husserl's thinking has affected the history of science quite significantly, aside from possible influences of his thought on analytic notions such as that of a constitution-system. Friedrich Steinle's contribution gives us a practical application of his ideas to a case of sedimentation in the experimental sciences, one that apparently runs against Husserl's assumption that it occurs only in the deductive-mathematical ones. For Steinle, there is more sedimentation in experimental science than Husserl envisaged—the converse of Ian Hacking's criticism, according to which there is less sedimentation in mathematics than he supposed. His case study concerns Charles Dufay's discovery of the bipolar nature of electricity.

According to Steinle, Dufay's research was directed more toward articulating *concepts* than it was toward *theories*. Dufay did not offer any explanatory theory to go along with his conceptual articulation; thus we are not dealing in his case with an episode in the deductive sciences. But Husserl's concept of sedimentation, while it was developed with an eye on the axiomatic, a priori sciences, is nevertheless just as applicable in cases such as Dufay's. For in the laboratory sciences, knowledge can sediment as materials and methods. In particular, the long sequence of conceptual articulation required by Dufay to attain his new concept of electricity eventually sedimented for his successors because they were able to purchase devices and textbooks that rendered the two electricities immediately accessible to them. Their unreflective use of these concepts and equipment both contained and concealed the results of Dufay's painstaking laboratory work.

Steinle agrees with Hans-Jörg Rheinberger in seeing parallels between Husserl's notion of sedimentation and Fleck's account of the origination (*Entstehung*) of scientific facts, according to which these bear traces of the now-hidden experimental and theoretical activity that generated them. The existence of such processes suggests to Steinle that science may not have the sedimentary structure Husserl ascribed to it. Instead of a successive layering of axiomatic systems, Steinle sees a haphazard structure reminiscent of a coral reef, whose living and sedimented parts are in flux, and which can only be understood by investigating its organic history.

The breadth of these essays is reflective of the scope of Husserl's *Crisis*, which may strike us as absurdly ambitious by present standards. Few philosophers today feel able, and more importantly, they do not feel compelled, to do philosophy on the scale that Husserl does in this book. There can be no doubt that this is one of the last works of philosophy that is free of the nagging self-doubt typical of much philosophy in the latter half of the twentieth century. The collapse of both structuralism and the project of unified science left both continental and analytic philosophy in similarly fragmented states. Perhaps, as Michael Hampe argues, Sellars was an exception to the rule, but it is fair to say that philosophy after the 1950s is chastened and modest in its ambitions, having largely retreated to the analysis of language. Husserl's book, on the other hand, seems like a remnant of classical German Idealism stranded in the twentieth century. It knows no restraint; and if it is despairing, it is emphatically not jaded.

But the *Crisis* still makes for instructive reading, perhaps for this very reason. Is not the process of dissolution and self-doubt that infects postwar philosophy something the *Crisis predicted*? Is it not a symptom of the crisis itself? Postmodernism began by explicitly *rejecting* Husserl's solutions, but that meant first accepting in large measure the diagnosis he had offered, in particular the account of sedimentation and axiomatic (written) systems. Similarly, analytic philosophy has conducted a long campaign to reduce the mental to the physical but has only recently come to acknowledge the tenacity of intentional and semantic concepts. The philosophy of mind operates in the very gap between life-world and science that Husserl so precisely surveyed. It lives, one might say, in a permanent crisis.

This is not to say that a modern-day reader will have an easy time extracting contemporary lessons from the *Crisis*. As with most late works, we are in the presence of a master, but one who is increasingly alone, who feels ever less of a need to explain or justify himself. That makes comprehension arduous. But the effort is well rewarded. Anyone who takes the time to work through his arguments stands to gain much from these night thoughts of Husserl.

Science and the Life-World

§ 1 Science, Intentionality, and Historical Background

David Woodruff Smith

Husserl's Philosophy of Science

When we think of philosophy of science, we think of Ernst Mach, Rudolf Carnap, Carl Hempel, Ernest Nagel, W. V. Quine, Thomas Kuhn, Patrick Suppes, and others. We do not at first think of Edmund Husserl, noted for launching phenomenology as the reflective, first-person science of consciousness—in contrast with physics taken as the hypothetico-deductive, third-person science of nature at its most basic level. Yet Husserl explicitly addressed many issues in the foundations of mathematics, logic, and science at a time when the philosophy of science was taking shape around twentieth-century mathematical physics and mathematical logic. The term of art in Husserl's day and milieu was not *philosophy of science*, but *theory of science*, or *Wissenschaftslehre*. In his lifetime Husserl had direct contact with Weierstrass, Kronecker, Cantor, Frege, Hilbert, Schlick, and Carnap, and his work was known to Tarski and Gödel. That is to say, Husserl was hardly out of the loop of thinkers then working on the foundations of mathematics, logic, and science. Husserl was not of course an architect of relativity theory, non-Euclidian geometry, number theory, set theory, or quantifier logic. But he was ever expanding a wide-ranging *philosophy* that addressed all these "sciences" (in the wide sense of the term then in vogue) as well as psychology and phenomenology.

Husserl's first book was published in 1891 as *Philosophie der Arithmetik*. Partly due to Frege's (understandable) misreading, that book was viewed as a psychologistic reduction of numbers to merely psychological activities of grouping and counting. A more careful and subtle study finds Husserl carrying out an analysis of the concepts of number and totality, which intentionally represent numbers and totalities. (This sympathetic reading of Husserl's first book is outlined by Dallas Willard in his introduction to his English translation of Husserl's book, *Philosophy of Arithmetic*; see in particular pp. xxi–xxix.) In terms of Husserl's mature thought, he was aiming at an analysis of the *sense* (*Sinn*) of number and the coordinate *essence* (*Wesen*) of number, where a sense is an ideal intentional content and an essence is an ideal property. But these notions Husserl developed only later in his full theory of intentionality and its attendant ontology.

Husserl's first mature work was the *Logical Investigations*, first published in 1900–1901 in three volumes spanning some 1,000 pages. In that work Husserl developed a unified philosophical system of logic, semantics, ontology, phenomenology, and epistemology. As Husserl explained in the opening volume, titled "Prolegomena to Pure Logic," that system was inspired in good measure by Bernard Bolzano's 1837 groundbreaking opus *Theory of Science* (*Wissenschaftslehre*) (1972). Accordingly, Husserl was seeking inter alia a basic account of the logical and epistemological structure of any science. A crucial part of the *Investigations* was Husserl's detailed theory of intentionality, the foundation of phenomenology, which was followed by an extended analysis of evidence or "intuition" (*Anschauung*) in the theory of knowledge, starting with sense perception. Issues of logic, ontology, intentionality theory, and epistemology thus frame the background for Husserl's subsequent programmatic presentation of the discipline of phenomenology in *Ideas I* (1913).

Empirical science, from physics to psychology, was prominent in the horizon of Husserl's phenomenology (even in his technical sense of *horizon*). Phenomenology was to be sharply distinguished from empirical psychology, which would include today's cognitive science, because phenomenology seeks ideal structures of meaning (Sinn) in the content of experience, whereas empirical psychology seeks contingent patterns of mental activity and its causes, even as in today's functionalist models in cognitive science and cognitive neuroscience. Physics too lies beyond the scope of phenomenology because phenomenology (in one range of analysis) is to bracket the existence and the essence of physical nature (even, we note today, the essence of the neural activities on which our consciousness depends) in order precisely to shift our attention to the way we *expe-*

rience things in nature. Thus we shift our attention away from the objects of our consciousness (in everyday perception or indeed in the abstract thinking we might practice in doing mathematical physics), and we turn instead to our own forms of experience, and therewith to their contents or meanings. Armed with phenomenological analyses of perception, judgment, and so forth, we will however return to an analysis of the structure of knowledge, ranging from our familiar knowledge of everyday affairs to our collective scientific knowledge, as in mathematical physics. That is where the *Logical Investigations* ends: poised for more specific analyses of formations of knowledge, say, in the natural sciences. As we consider below, philosophy of science followed out such an analysis not least in the work of Rudolf Carnap, in ways surprisingly congenial to Husserl's intentionality based "theory of science." Keep in mind that Husserl's conception of intentionality and phenomenology is framed by his wider theory of science, *überhaupt*.

Husserl's last wave of work, during the period 1935–38, was collected in the posthumous book called *The Crisis of European Sciences and Transcendental Phenomenology*. The intellectual and spiritual crisis Husserl saw resides in the way in which mathematical physics has lost touch with everyday human experience. Galileo's vision of science began the ideal of a "mathematization" of nature, which in the early twentieth century produced first general relativity theory and then quantum theory. As widely discussed, both relativity theory and quantum theory seem incompatible with our *Lebenswelt* (life-world) experience and our everyday knowledge of the world around us—even though science begins with everyday perceptual observation and proceeds with work in the laboratory, at the computer screen, and so forth. Husserl addressed this problem concerning the foundations of twentieth-century physics in a "transcendental" critique of mathematical physics.

As Husserl's analysis of science unfolded from his early work in the *Logical Investigations* to his late work in the *Crisis*, we find a strong tension. The early vision is full of hope for the model of any science as a formal, ideally mathematical theory of a given domain supported by evident experience or intuition of an appropriate type. But the late vision finds a crisis, as mathematical physics no longer seems tied to human experience, and we can no longer really understand what mathematical physics—Einsteinian physics in particular—says about the world around us.

Our present task is to explore certain issues in a Husserlian philosophy of science, looking to its foundation in a Husserlian theory of intentionality and attendant ontology. We will be pressing three themes further than Husserl did. The first theme is the indexical structure of perception (seeing "this such-and-such"), which is the form of experience that serves as the evidential base of any empirical science. The second theme is the "global" indexicality of our concepts of things in nature (such as the concept *tree*), concepts that by their very meaning represent things in our surrounding world, or *Umwelt*. The third theme is the ontological dependence of our theories, even in mathematical physics, on historical human intentional activities.

The results of this exploration allow a new view of the formal and material aspects of science, which we shall bring into relief by drawing on Michael Friedman's Kantian-Carnapian-Kuhnian account of science.

Husserl's Unified Theory: A Theory of Theory, Essence, Meaning, Part/Whole, Intentionality, Evidence—and Thus Empirical Science

Husserl's *Logical Investigations* (1900–1901) unfolds a unified system of logic, ontology, phenomenology, and epistemology: defining a framework within which we may outline a Husserlian philosophy of science. The *Investigations* includes seven interrelated studies. The Prolegomena outlines a theory of "pure logic," which prefigures the development of semantics or metalogic by Carnap, Tarski, and others. Investigation I offers a theory of linguistic expression, meaning, and reference (with some similarity to Frege's). Investigation II offers a theory of universals, or "ideal species" or essences (Wesen), combining elements of the Platonic and Aristotelian views of universals. Investigation III is a theory of part and whole, including the notion of "moment," or dependent part. Investigation IV applies that part/whole ontology to ideal meanings in a theory of "grammar." Investigation V is a long presentation of Husserl's groundbreaking theory of intentionality, carefully distinguishing act, content, and object of consciousness. That theory of intentionality is the foundation of Husserl's conception of phenomenology. Finally, Investigation VI develops a phenomenological theory of knowledge, featuring the character of evidence in intuition, including visual perception. What is remarkable, and too little appreciated, is the tightly knit unity of Husserl's system. In my view Husserl joins Aristotle and Kant among the most systematic of philosophers.[1]

Looking to philosophy of science, I should like to emphasize three parts of Husserl's system: his theory of theories, his theory of intentionality, and his theory of evidence. Husserl's system as a whole is a wide-ranging theory that includes a theory of what it is to be a theory. That is, Husserl's theory of almost everything includes a metatheory that applies to the overall theory. And the metatheory is itself a *part* of the theory, in the sense explicated in Husserl's theory of parts/wholes. (How does this play today in light of the results in metalogic that followed in Husserl's wake?)

Husserl's theory of theory (Prolegomena) holds that a *theory* is a system of propositions, which are ideal *meanings* (compare Bolzano's notion of *Satz an sich* and Frege's notion of Sinn or *Gedanke*). A deductive theory includes deductive consequences of propositions in the theory; an inductive theory includes propositions made probable ("motivated") by other propositions in the theory. Each theory forms a semantically coherent group of propositions that characterize a *domain,* or field of knowledge. The propositions in a theory develop "laws of essence" about the domain of the theory, that is, they specify the *essence* (Wesen), or properties, of objects in that domain. All this should sound familiar to us today. But notice that the elements in a theory are propositions, ideal propositional meanings (Sinne), rather than sentences in a language (as subsequent logicians, such as Carnap, Quine, and others, assume). And bear in mind that in Husserl's mature ontology, essences and meanings are categorially distinct. Meanings are ideal intentional contents that semantically represent objects and their properties or essences. As an approximation, essences are like Aristotelian universals, while meanings are like Fregean Sinne.

If a theory is a system of propositions, what kind of entities are propositions for Husserl? Propositions are meanings that are propositional in form (expressible by a complete declarative sentence). But meanings are, for Husserl, ideal *contents* of intentional acts of consciousness. In the *Investigations,* Husserl identified ideal meanings with ideal species of intentional acts: A meaning is the ideal species of an experience of thinking or seeing or whatever. So a proposition is the ideal species or type of thinking that such-and-such is the case. By the time Husserl wrote *Ideas I,* however, he had come to hold instead that meanings are their own kind of ideal entities: not species of intentional act, but meanings sui generis, like Fregean Sinne, if you will.

What makes a theory *true*, for Husserl, is its correctly representing the essence of objects in its domain. But truth is distinct from evidence. What makes a theory count as *knowledge* of its domain is the evidence supporting it. In Husserl's phenomenological theory of knowledge (Investigation VI), evidence consists in the phenomenological character of "evidence" (*Evidenz*) or "intuitive fullness" or "fulfillment." An act of intuition (Anschauung) is an intentional act of consciousness that has the character of intuitiveness, that is, intuitive fullness or evidentness. When I see that red ball, part of the way the object is experienced is with a character of sensory evidence. I do not merely think of a red ball, I see—or, some insist, seem to see—a red ball. If I then judge visually that it is a red ball, my judgment has a meaning supported by the character of intuitive evidence in my act of so judging.

Husserl distinguishes some three types of intuition. Sensory or empirical intuition is found in sensory perception. Essential intuition (*Wesenschau*) consists in coming to "see" something about an essence or ideal species of object, for instance, that a triangle has angles totaling 180 degrees, coming to see this perhaps through deduction or through eidetic variation. Transcendental phenomenological intuition consists in grasping the structure of an experience as lived, including its ideal intentional content or meaning, perhaps through bracketing or epoché. The techniques for achieving these types of intuition are very different, and beyond the scope of our discussion. I wish only to stress that there are different types of evident or intuitive experience or judgment, achieved in different ways.

A distinguishing feature of Husserl's theory of theory, and thus of knowledge, including science, is his view that a theory is a form of intentionality, a system of propositions that serve as contents of judgments about a given domain of objects. In *Formal and Transcendental Logic* (1929), Husserl treats formal logic, where a theory is a formal or symbolic deductive system, as requiring a foundation in transcendental logic, by which Husserl means formal logic grounded in intentionality. As we might say today, a formal theory (a system of symbols, rules of inference, axiomatic sentences) is accompanied by a semantics, a system of assignments of meanings to relevant expressions in the formal theory. But meanings, for Husserl, are ideal contents of intentional acts of thought, perception, and so forth. Thus we find the later Husserl amplifying the early Husserl of the *Logical Investigations*. Husserl's theory of scientific knowledge flows from this intentionalist account of a theory supported by appropriate evidence in perceptual experience.

In Husserl's theory of knowledge, we may say that empirical knowledge consists in forming empirical judgments with appropriate sensory intuitive evidence. As Quine would say, empirical knowledge begins with observation, with sensory perception of objects. Husserl would specify then that an observation consists in a perceptual judgment, an intentional act with a certain propositional content and a certain character of sensory evidence.

When does empirical knowledge develop into what we call a *science* like physics? In Husserl's scheme, a science like physics is an accumulation of empirical knowledge produced by a community of intentional subjects building on the work of others. A *scientific theory*, we may say, is thus a theory—a system of propositions (ideal meanings) about a domain of objects—that is supported by evidence (intuitive fulfillment) in observations (perceptual experiences) and put forth in judgments performed by individuals working collectively in the relevant scientific community.

In the *Crisis* (1935–38), drawing on results as early as *Ideas II* (1912), Husserl characterized what he called the *sedimentation* of knowledge. Traditional beliefs are sedimented propositions that we typically share while having little if any explicit awareness of these belief-contents. These may be called background beliefs. But mathematical knowledge too is sedimented. Husserl focuses on Euclidian geometry, the system of propositions we have inherited and developed since Euclid. After Galileo, Husserl emphasizes, modern European science developed a vision of the mathematization of nature. That is, Husserl argues at length, physics after Newton came to lay down laws of nature—laws of motion and gravity—that are expressed in mathematical language, starting with the calculus. The calculus itself rests on Cartesian analytic geometry, which uses algebra to reformulate principles of Euclidian geometry. All these branches of mathematics are themselves accumulations of results produced by individual mathematicians working within the extended historical tradition and community of mathematics. For Husserl, then, mathematical physics is a doubly sedimented accumulation of knowledge. Thus we distinguish within Newtonian physics two bodies of theory, or simply two theories: There is the mathematical theory of continuous functional relationships that we call the calculus, and there is the physical theory of continuous motion in space-time that we call Newtonian mathematical physics. We shall return to the difference in status between these two bodies of theory. For now, though, I want to emphasize the sedimentary feature shared by both the mathematical theory and the physical theory.

For Husserl, we have stressed, a theory is a system of propositions, which are ideal meanings and which serve as contents of intentional acts of consciousness. Some readers may tend to think of ideal meanings in a Platonistic way, as floating above the world in a heaven of ideal entities. This caricature is deeply misleading. Instead, we should think of ideal meanings—and ideal essences too—as part of the world. Indeed, meanings are tied into the wider world in very specific ways, not least in everyday perception and in scientific theory. I want to focus now on just those ways, in exploring the foundations of a Husserlian intentionality based philosophy of science. We shall be adapting and extending Husserlian theory as we go.

Intentionality, Indexicality, Global Indexicality, and Historicality in Husserl's Theory of Science

In a Husserlian theory of scientific knowledge, ideal meanings are tied into concrete events in several different ways: as contents of acts of consciousness, as indexical meanings in contextualized perceptual experience, as globally indexical meanings in empirical judgments about "this world," and as theoretical propositions developed in a historical scientific community.

On Husserl's general theory of intentionality, every act of consciousness is directed toward an object (of appropriate type) through a particular meaning that semantically prescribes or is satisfied by that object (if such object exists). By hypothesis, then, every act of perception, thought, imagination, and so forth has—is correlated with—a certain meaning. As McIntyre and I put it, an intentional act or experience *entertains* a meaning. The act is a concrete event or state "performed" by a given person or subject on a given occasion, and the meaning it entertains is an ideal meaning that can be entertained by other acts as well (so the correlation of acts to meanings is many to one). That relation of entertaining is a primitive in the Husserlian theory. I shall simply assume that relation here as we move on to the specific role of meanings in scientific knowledge.

Empirical theory is grounded, evidentially, in sensory perception. Husserl's phenomenology studied perception at length. Classical empiricism assumed that we fundamentally perceive sense data—patches of color, patterns of sound, and so forth. However, Husserl joins Kant in holding that perception always (or at least normally) involves conceptualization as well as sensation: I see a red, round tomato, not a red, round sense datum. As Husserl puts it, my visual experience is a whole containing two dependent parts ("moments"): the sensory *hyle* (sensations or sensory "matter") and the interpretive *noesis* (apprehension giving "form") (*Ideas I*, § 84). The whole experience entertains a special type of meaning, what Husserl calls an "essentially occasional" meaning. I'll call it an essentially *indexical* meaning (as the term indexical has become entrenched in recent decades). For example, when I see an object before me, we may capture the phenomenological structure of the experience by the form of phenomenological description: "I see this red, round beefsteak tomato."

Well ahead of his time, Husserl began the outlines of a semantics of indexical words: The word *this* as uttered on a particular occasion refers to an object of the speaker's perception on that occasion. (Seventy years later, David Kaplan started his semantics of demonstratives with distinctions similar to Husserl's, while Hilary Putnam drove home the point of indexicals with his "Twin Earth" scenario, even as Husserl had recounted a form of Twin Earth scenario sixty-five years earlier in his 1908 lectures on meaning.) With such a semantics in mind, I propose (as detailed in Smith, 1989) an explication of the *intentional force* of the indexical meaning in a perception:

> The indexical meaning "this" entertained in a visual experience in a given context (or occasion) prescribes or is satisfied by an object in that context if and only if that object is appropriately visually before the subject of the experience on the given occasion.

Thus, the content "this red, round beefsteak tomato" in my current visual experience semantically prescribes or is satisfied by a certain object just in case that object is red and round and a beefsteak tomato and is visually before me on the occasion of my perception—that is, the object is at a certain place before me and is playing an appropriate role in partly causing my experience as light deflected from the object is affecting my eyes.

In this way perception is intentionally tied into the immediate context or occasion of perception. On the present theory of indexical intentionality, the indexical meaning of "this" entertained in a perceptual experience appeals semantically to the *context* in which the perception occurs. The meaning is ideal, but its intentional force is directly tied into the world in the concrete context of experience. On a different occasion, the meaning of "this" will prescribe a different object, visually before the subject on that occasion, in that context. Thus it is only when entertained in a concrete experience in a concrete context that the indexical meaning of "this" has any intentional force at all: Its semantic or intentional force is essentially context-dependent.

Now empirical theoretical concepts are also—in a partly similar way—tied into the wider context of our earthly human experience. For instance, the meaning of *gum tree*, or the more precise concept *Eucalyptus globulus*, as refined in botanical theory, prescribes or is satisfied by a species of tree found in Australia and imported to California, a tree with tall and swaying trunk, strips of peeling bark, and fragrant oil-seeping leaves. Where "this" is an *individual concept* that appeals to the context of perception, *Eucalyptus* is a *general concept* that appeals to the geographic context of the Australian continent where this species of tree evolved. Moreover, where "this" is an *observational* meaning, operative in perceptual observation, *Eucalyptus globulus* is a *theoretical* meaning, that is, a concept that is part of a wider biological theory of plant life on Earth. Indeed, this theoretical concept is part of a scientific theory that has been developed in a long-running tradition of scientific observation and theory concerned with relevant parts of the world we find ourselves in—as practicing biologists or as lay people learning bits and pieces of biology from the experts. We may say, then, that theoretical concepts or meanings in natural science have a *global indexicality*. If you wish, our theoretical concepts and propositions are *global* as opposed to strictly indexical meanings: Their meaning appeals intentionally or semantically to our context of living on *this* "globe," our planet Earth, amid the rock, atmosphere, flora, fauna, primates, and human culture that surround us. (Globalism takes many forms.)

We can draw such a picture out of Husserl's rather abstract account of the "constitution" of "things" in nature in *Ideas I*, § 149–51. For Husserl, in his revamping of the Kantian notion of "constitutive rules" governing phenomena, an object is *constituted* in a "manifold" (*Mannigfaltigkeit*) of meanings, or *noemata*, insofar as that same object is represented in different ways, as having different aspects, including aspects perceived from different perspectives. If Husserl is read as a genuine idealist, then this structure of constitution literally brings the object into being. However, if Husserl is read as a realist or rather as a perspectivist (as I read him for the most part), then constitution is a matter of intentional or semantic representation. Corresponding to every object, on this view, there is a structure or manifold—a "horizon"—of noematic meanings that represent the object from various perspectives in different but "harmonious" ways. (See Smith, 2006, on constitution, horizon, and manifold, and Smith and McIntyre, 1982, on horizon.) And given Husserl's conception of pure logic in the *Logical Investigations*, there is a parallel structure or manifold of properties in the essence of the object. The *ontological* constitution of the object consists in this manifold of properties, corresponding to the *phenomenological* constitution of the object through the manifold of meanings representing the object. Now in the constitution of a thing (*Ding*) in nature, Husserl holds, there are well-defined "strata" of meanings: first the meanings that represent temporal properties of the thing, then meanings that represent spatial properties of the thing, then meanings that represent causal relations of the thing in nature, then meanings that represent the thing's being experienced by a subject in intentional relationships, and then meanings that represent the thing's intersubjective experienceability and therewith also its value to a community.

Husserl summarizes these levels of constitution in the essence of a thing: "The thing in its ideal essence presents itself as *res temporalis*, . . . as *res extensa*, . . . as *res materialis*, [as] it is a *substantial* unity, and as such the unity of *causal connections*, . . . '*and so forth*' . . ." (*Ideas I*, § 149). As meaning and essence are distinct, the stratified structure of the essence of a natural thing is *represented* intentionally or semantically by a stratified structure of the meaning "natural thing." Now, the stratified essence of a thing in nature is defined by a "material ontology" of the region of nature. And this region encompasses things in the spatio-temporal-causal order within which we happen to find ourselves in our range of experience as psychophysical animals here on planet Earth. In other words, we generally encounter and deal with things in nature within a global context, or Umwelt. As Husserl says at the beginning of the long phenomenological investigation prior to this analysis: His changing "spontaneities of consciousness" are related to "this world, *the world in which I find myself and which is likewise my surrounding world* [*Umwelt*]" (*Ideas I*, § 28). This world is the world of nature, but also the arithmetic world with numerical quantities and the intersubjective world with other people and even values. This Umwelt of nature cum consciousness cum culture becomes the Lebenswelt of Husserl's late work, the *Crisis*. (See Smith, 1995, for a reconstruction of Husserl's account of formal and material ontology, including the regional essences of nature, consciousness, and culture, or *Geist*.)

Bear in mind that there is only one world, although the same object can have different properties or essences and can be intended through different meanings and from different perspectives. When I gesture with my hand, I take the hand as part of my lived body (*Leib*) and use it as part of the everyday practice of gesturing in the Lebenswelt, part of my everyday cultural world. By contrast, when I inspect my hand after an injury, wondering how deep is this cut, I take the hand as part of this physical body (*Körper*), this organism in nature, part of the natural world. I might go on to wonder about the DNA in this tissue, part of the world of biological nature. Thus where Husserl distinguishes the Lebenswelt, or life-world, from the natural world, we should understand this as a distinction at the level of meaning, a distinction between different ways of intending objects in the world as having different types of essence, a distinction between objects *as* experienced or meant in various ways. The Lebenswelt is the world *as* I experience it in everyday life, the world *as* I find it around me, my Umwelt. Furthermore, my Umwelt is distinct from your Umwelt, and my Lebenswelt is distinct from your Lebenswelt, not because there are two distinct worlds (mine and yours), but because my indexical perspective on the world is distinct from yours. That is, you and I experience the same world but through different indexical meanings, say, as you and I see the same tree, "this" tree, from different perspectives relative to your eyes and my eyes. (See Smith, 2006, especially chapters 5 and 6, on Umwelt and Lebenswelt in Husserl's phenomenology.)

On the view I am sketching, our theoretical meanings or concepts are global, or globally indexical, in that they represent objects in "this world," "my Umwelt," which is in its more basic strata, a world of things in nature. In the posthumous volume *Experience and Judgment*, Husserl explicitly recognizes this global character as he marks

> *the difference between free idealities* (such as logico-mathematical systems and pure essential structures of every kind) and *bound idealities*, which in their being-sense carry reality with them and hence belong to the real world. All reality is here led back to spatiotemporality as the form of the individual. But originally, reality belongs to nature; the world as the world of realities receives its individuality from nature as its lowest stratum. . . . Bound realities are bound to Earth, to Mars, to particular territories, etc. But free idealities are in fact also mundane: [bound] by their historical and territorial occurrence, their "being discovered," and so on. (*Experience and Judgment*, p. 267)

So *bound* means bound to a territory such as Earth or Mars. However, both ideal essences and ideal meanings may be bound to Earth, and they are bound in different ways. In my example above, the *species Eucalyptus globulus* is an ideal essence bound by nature to Earth, originally to Australia, where it evolved. But the *concept Eucalyptus globulus* is also a bound ideality: It is an ideal meaning bound by semantics, by global indexicality, to its assumed territory of application here on planet Earth, first to Australia and later also to California. If there is a Twin Earth in a distant galaxy, our use of the concept *Eucalyptus globulus* in our scientific thought and discourse here on planet Earth does not refer to any similar species on Twin Earth. In our thoughts, the concept *Eucalyptus globulus* is semantically bound to its assumed earthly habitat.

Husserl remarks above that even "free" idealities, nature-free essences such as those defined by pure mathematics and logic, are bound by their "being discovered." Here is a third and very different way in which ideal entities are tied into the historical context of our theorizing: not by semantically or intentionally appealing to "this" object or "this *Umwelt*," but by being discovered or known or, most generally, "intended" in certain historically situated activities of consciousness performed by individuals in a community or tradition, as by practicing scientists or mathematicians. As Husserl stresses in the *Crisis*, geometry developed with a history since its "origin" in Euclid. Thus the essences of lines, angles, and circles are *historically* tied by epistemic or intentional relations to the discoveries of Euclid and other mathematicians here on Earth—even though these essences are not nature bound in the way that the species *Eucalyptus globulus* is bound into earthly botanic life. Furthermore, the *concepts line, angle,* and *circle*—as distinct from the essences they represent—have been developed through the historical activities of mathematicians in human history here on Earth. These concepts are not what we call indexical or even globally indexical concepts, yet they are *historically* tied to their development, even as we use them in our current geometrical thoughts. Husserl from time to time cites non-Euclidian geometries, underscoring the way in which these mathematical systems of propositions (ideal meanings) are free of their application to nature, yet these propositions are historically bound to their development.

Nature-bound essences too are *historically* tied to their scientific discovery. However, this historical binding is different from the ways in which natural essences are bound into nature. The essence of electrons is *naturally* bound into our universe, as the essence of Eucalyptus trees is naturally bound into our Earth. But these natural essences are also *historically* bound to our earthly domain insofar as we discover them through our scientific practices of collaborative observation, hypothesis formation, and testing in various periods of human history here on planet Earth. In this way, the essences of Eucalyptus trees, gravitational forces, electrons, and strings are historically tied to their discovery in historical practices of natural science. Now, the *concepts* or meanings *gravity, electron,* and *string* have been not so much discovered as developed. And through these meanings the corresponding essences are *constituted,* in Husserl's technical sense. In physics, moreover, we make use of both mathematical meanings, such as *number, set,* and *vector,* and empirical meanings, such as *gravity, electron,* and *string.* These mathematical and empirical meanings have been developed in distinguishable activities of pure mathematics and empirical mathematical physics. The intellectual practices in which these meanings have been developed are historical processes. Accordingly, in the *Crisis* Husserl addresses both the origin of geometry in Euclid and the mathematization of nature since Newton in specific historical processes. The point to emphasize here is the way in which both mathematical theory (geometry, number theory, and set theory) and physical theory (Newton's theory of gravity, electromagnetic theory, relativity theory, and quantum theory) are historically tied into intentional activities of perceptual observation and theorizing.

For Husserl, then, meanings are ideal yet historical entities. (Not your standard Platonism!) Accordingly, in the practice of science we do commerce with concepts and propositions that carry their history with them, that is, their historical ties to intentional acts of consciousness performed by other people at other times and places, activities in which these scientific concepts and propositions are developed as such.

Our account of indexicality, global indexicality, and historicality helps to resolve some of Husserl's worries in the *Crisis*, to the effect that mathematical physics has lost touch with our everyday Lebenswelt experience and its Umwelt. We should not view our mathematical physical theory as a "free" ideality, a pure and free mathematical theory that describes only a free mathematical structure. Instead, we should see our mathematical physical theory as "bound" to our everyday experience and our surrounding world, from our own laboratory to our own planet Earth to our own universe with its Big Bang origin. Both observational perceptual meanings (*this tree*) and theoretical physical meanings (electron, string, DNA, *Eucalyptus globulus*) are bound to the context of our experience insofar as their intentional force reaches into the Umwelt of our observing and theorizing in our intentional constitution of nature ranging from our *indexical* everyday experience through our *global* conceptualizations in mathematical physics. And furthermore these meanings are *historically* bound to the collective intentional activities in which the relevant scientific theories have developed and are continuing to develop even as we carry on with historically developed and "sedimented" scientific concepts and theories.

Husserl vis-à-vis Carnap's Constitution Theory of Empirical Science

The structure of logical representation and empirical evidence is central to Husserl's *Logical Investigations*, to his inaugural conception of phenomenology therein, and to his mature conception of *constitution* from *Ideas I* to the *Crisis*. The same structure of representation and evidence figured centrally in the developing genre of twentieth-century philosophy of science. The most rigorously formulated theory of how sensory evidence and logical structure join together in a theory of knowledge, including physics, was surely Rudolf Carnap's *Der Logische Aufbau der*

Welt (1928), translated as *The Logical Structure of the World* (2003). Today's philosophers of science do not usually associate Carnap's project with Husserl's. Yet Carnap had attended Husserl's lectures in 1924–25, during the years he wrote the *Aufbau*, and there are important conceptual connections between Carnap's model of empirical knowledge and Husserl's. Both develop a logical structure of sentences or propositions supported by empirical evidence. Carnap casts his system in a language shaped by twentieth-century logic, where the language as it were embodies the knowledge codified in the system. Husserl, however, casts his system in a framework of ideal propositional meanings expressible in principle by an appropriate language, which could be a system of quantifier-predicate logic of the kind Carnap employs. Carnap grounds his system in the evidence of sensory elementary experiences, in part reflecting a classically empiricist model. Husserl, however, grounds his system in evident perceptual experience, or sensory intuition (Anschauung), analyzed as a fusion of sensory and conceptual "moments," coming closer to a Kantian than a Humean view of the ultimate evidence in sensory perception. In any event, what I want to stress is the role of intentionality or intentional constitution in empirical knowledge.

Carnap's original title for the treatise that became the *Aufbau* was *Entwurf einer Konstitutionstheorie der Erkenntnisgegenstände*, or *Outline of a Constitution Theory of the Objects of Knowledge* (see the discussion in Friedman, 2000, 70). The German term *Aufbau* connotes either the structure built or the process of building, just as the English term *building* can be used either as a noun ("the building is by the river") or as a gerund ("the building of the cathedral took centuries"). There is a similar process/product ambiguity in the term *constitution*. The theory of constitution is a familiar part of Husserlian phenomenology and grounded in his theory of intentionality: Objects of certain types are "constituted" in intentional acts insofar as they are represented or "intended" through appropriately complex systems or manifolds of ideal meaning. The system of meanings through which an object is constituted may include, ideally, a system of propositions that come together to form a theory about such objects, that is, according to Husserl's theory of theories in the Prolegomena of the *Logical Investigations*. Carnap's construction (Aufbau) of objects represented in a language in a piece of empirical theory is a detailed exercise in such a constitution (*Konstitution*) theory, where the medium of constitution is a language of formal or symbolic logic.[2]

Husserl's term constitution entered the philosophical lexicon with an eye to Kant, of course. Kant's doctrine of transcendental idealism may be that the human mind quite literally constructs the phenomenal world, including the world of Newtonian physics. That interpretation of Kant is a useful foil, even if it is not the only reading of Kantian doctrine. Indeed, an idealist or antirealist neo-Kantian view of knowledge was in the air in German philosophy, in the Marburg School, when Carnap wrote the *Aufbau*. Moreover, Carnap and other logical positivists were also influenced partly by the conventionalism of Poincaré's view of mathematics. (See Friedman, 1999, chapters 3, 5, and 6, on these lines of influence.) Scientific realism, espoused later by Wilfrid Sellars and many others, allows an alternative Kantian doctrine whereby our best science gives us a scientific image of the world in itself, while our everyday experience presents the "manifest" image of things as they appear to our senses filtered by our everyday concepts (on Sellars, see Michael Hampe's contribution to this volume).

As we saw, a Husserlian theory or philosophy of science will be grounded in the Husserlian theory of intentionality, as applied to scientific knowledge. Now Husserl's theory of intentional constitution is basic to his phenomenology and his theory of knowledge, and in *Ideas I* (1913) he adopts Kant's idiom of "transcendental idealism" while speaking of constitution. But it remains a point of scholarly dispute what constitution amounts to for Husserl: Does consciousness create or construct the objects of intentional experience? A realist, perspectivist interpretation says, no, an object is constituted in intentional experience insofar as it is represented through ideal meanings that cohere in meaningful patterns (defining the "horizon" of the object as experienced).[3]

Carnap's constitution theory of knowledge in the *Aufbau*, then, reflects the central issues detailed in Husserl's philosophical system laid out in the *Logical Investigations* and amplified in his later works. Following Husserl, Carnap adapts the term constitution within the context of neo-Kantian debates in German philosophy in the early decades of the twentieth century. But Carnap holds back from neo-Kantian idealism and also realism:

> Are the constituted structures "generated in thought," as the [neo-Kantian] Marburg School teaches, or "only recognized" by thought, as realism asserts? Constitution theory employs a neutral language; according to it the structures

> are neither "generated" nor "recognized," but rather "*constituted*"; and it is already here to be expressly emphasized that this word "constitution" is always meant completely neutrally.[4]

Carnap's stance of neutrality here is equivalent to Husserl's stance of epoché, or bracketing the question of the existence (or nonexistence or mind-dependence) of the objects constituted in acts of consciousness. Remember that the term *methodological solipsism* was coined in the *Aufbau*, and Carnap's initial stages in his "construction" are "autopsychological." (We today hear Jerry Fodor's talk of methodological solipsism as a research strategy in cognitive science [Fodor, 1982], and we hear Daniel Dennett's talk of "autophenomenology" as something to be replaced with "heterophenomenology" [Dennett, 1991].)

Carnap's logical structures are *formal* as opposed to material structures. Indeed, Carnap stresses that objects are constituted only through logically formed definite descriptions, strictly "syntactic" structures, which represent only structural features of the object and its relations to other objects.[5] These formal structures require a different semantics than descriptions if we bring in indexical, global, and historical factors as above; indeed, with Husserl's conception of intentionality, we advance explicitly from syntax to semantics (for thought as well as language). Setting aside indexicality and so forth, though, Carnap's constitution theory of empirical knowledge echoes Husserl's conception of pure logic as it expanded into a phenomenological theory of knowledge in *Logical Investigations.* The basic model of knowledge in this Husserlian system is an intentionality based theory of knowledge. By the time Husserl wrote *Ideas I,* he had folded in a somewhat Kantian notion of constitution, and Carnap's logical empiricist—better, logical phenomenological—model of empirical knowledge follows suit.

Husserl vis-à-vis Quine's Web Model of Empirical Theory

W. V. Quine has stressed the different roles of observations and explanatory hypotheses in an empirical theory, whether a part of our everyday belief system or a specialized science like physics (see Quine and Ullian, 1978). Wary of entities called meanings, Quine treats a theory as a logically ordered system of sentences that formulate a system of belief or knowledge. Observation sentences report the observations that form the empirical or evidential base of the theory, while theoretical sentences formulate the higher reaches of belief or knowledge articulated in the theory. Quine treats the overall theory as a "web" of belief put to language, where the observation sentences mark the empirically fixed perimeter of the web and the theoretical sentences are the center of the web. For Quine, logic and set theory form the innermost part of the web, as it were the most purely theoretical part of our belief system. This overall picture of empirical science is broadly at home in Husserl's theory of knowledge. However, for Husserl, we stressed, a scientific theory is a system of propositions (ideal meanings) expressible by sentences in a Quinean linguistic formulation of the theory. That said, we distinguish observational propositions from theoretical propositions. Within a Husserlian philosophy of science, I want to note the role of logical and mathematical propositions at the center of the web structure of an empirical theory.

Quine's familiar critique of Carnap features the breakdown of the analytic/synthetic distinction: thus Quine's holistic web model (see Quine, 1980). Yet Quine preserves a distinction between logical and nonlogical truth, the same distinction Bolzano drew. Now Husserl ramifies Bolzano's distinction. In his account of pure logic (in the Prolegomena of the *Logical Investigations*), Husserl distinguishes the "formal" from the "material" at three levels: language, meaning, and reality. Carnap and then Quine distinguish logical and nonlogical sentences or truths, where logical truths are sentences true by virtue of logical form alone. Husserl, however, applies the form/content distinction also at the level of the objects represented by sentences or propositions. For Husserl, formal ontology posits formal categories, including those of Individual, Property, State of Affairs, Number, Set, and so forth. These categories apply invariantly to any material domain or "region" such as nature—or within nature, the domains of the physical, the biological, the psychological, and so forth. In Husserl's scheme, material ontologies address the material essences of things physical, biological, and so forth, while formal ontology addresses the formal essences that apply invariantly to things with those material essences.

In Quine's web model, logical theory lies at the very center of our overall web of theory. Set theory and mathematical theory are, for Quine, the next ring of theory. Then, we may say, comes the ring of physical theory, from physics through chemistry to biology to neuroscience and on to empirical psychology. Then, if we nod to Husserl, comes the wide ring of everyday theory about our Lebenswelt. And finally at the perimeter of the web come our perceptual observations. This web structure can be explicated in some detail, it should now be clear, within a Husserlian framework, as the structure of the web unfolds the constitution of the world as we know it in our overall theory. We recall the indexical, global, and historical characters of the various meanings within the web of theory.

Within such a web model of knowledge, the outstanding issue between a Quine, a Carnap, and a Husserl is precisely how flexible are the different parts of the web, from the sensory to the logical elements, and how malleable each is in light of the others. The rationalist stresses the logical; the empiricist, the sensory. The naturalist places knowledge cum intentionality in a framework of nature, while the transcendentalist places nature in a framework of intentionality cum constitution. It should be clear that the Husserlian model of knowledge we have been exploring is not easily classified in these traditional terms.

Husserl vis-à-vis Friedman's Theory of the Relativized A Priori in Physics

Michael Friedman's recent analysis of science (1999, 2000, 2001) adapts a Kantian theory of knowledge to the developments of twentieth-century physics in light of Carnap's constitution theory together with Kuhn's theory of scientific revolutions. Friedman's analysis can be explicated in instructive ways that extend Husserl's theory of science along lines we have been developing above. Indeed, Friedman's account draws on Carnap's work on geometry and its role in relativity theory, which, Friedman shows, draws on Husserl's philosophy.

Friedman summarizes Kant's theory of knowledge in physics as follows:

> Kant's analysis of scientific knowledge—as articulated especially in his *Metaphysical Foundations of Natural Science* of 1786—is based on a sharp distinction between "pure" and "empirical" parts. The pure part of scientific knowledge consists of physical geometry (which, for Kant, is of course necessarily Euclidian geometry), more generally, the totality of applied mathematics

> presupposed by Newtonian physics (viz., classical analysis), Galilean kinematics (the classical velocity addition law), and the Newtonian laws of motion. In short, the entire spatiotemporal framework of Newtonian physics—what we now call the structure of Newtonian space-time—belongs to the pure part of natural science. The empirical part then consists of specific laws of nature formulated within this antecedently presupposed framework: for example, and especially, the law of universal gravitation and, more generally, the various specific force laws that can be formulated in the context of the Newtonian laws of motion. (1999, 59)

For this Kantian view of Newtonian physical theory then, Euclidean physical geometry of space (or space-time) is a "pure" theory of space presupposed by the "empirical" theory of gravity and motion in space. The pure part of the Newtonian theory is a priori and presupposed by the empirical part of the theory, which follows on observation. Moreover, the geometry in the physics is "constitutive" of the form of space. Nodding to Husserl (and Carnap), let us just say that the geometry "constitutes" the form of space. The laws of motion then describe movement in such a space. Importantly, as Friedman specifies, we are talking about applied, thus physical, geometry. The pure part of the physical theory is not pure as opposed to applied mathematics; rather, it is pure of empirical observation, that is, a priori, and it is presupposed by and so logically and epistemically prior to the empirical laws of motion.

In the middle of the nineteenth century, Riemann and other mathematicians began developing non-Euclidean geometries, pure mathematical theories waiting to be applied where appropriate. By 1915 Einstein applied a form of non-Euclidean geometry to space-time in his general relativity theory (Friedman, 1999, 44ff.). On Friedman's Kantian account then, Einsteinian general relativity theory consists of a pure and an empirical part. The pure part of the theory is the non-Euclidean geometry of space-time, and presupposing that geometry is the empirical theory of gravitation (and so forth). The new physical geometry is a new constitutive theory of space-time, and the new theory of gravitation describes movement following the curvature of space-time. (Husserl interacted with related foundational issues of mathematics and space-time: See Smith, 2006, on Husserl's wide use of his notion of manifold in regard to the structure of space and time as experienced and the structure of consciousness and its constitution of the world.)

Observing this Kuhnian revolution in physics, Friedman proposes a conception of *relativized a priori* theory that constitutes the form of space or space-time. Where Kant assumed that Euclidian geometry is our fixed a priori synthetic theory of space, the development and application of non-Euclidian geometries shows that a given physical geometry is a priori only within or relative to the physics that presupposes this geometry (see Friedman, 2001, 71–93). In the Kantian idiom Friedman employs, the applied geometry presupposed in the physical theory, Newtonian or Einsteinian, is a "transcendental condition of the possibility" of the empirical theory of motion (and so forth).

A perhaps stronger view presses the ontological priority of the mathematics used in twentieth-century physical theory. Luciano Boi stresses the "great significance of space-time geometry in *predetermining* the laws which are supposed to govern the behavior of matter, and further to support the thesis that matter itself can be built from geometry, in the sense that particles of matter as well as the other forces of nature emerges [sic] in the same way that gravity emerges from geometry" (2004, 429, emphasis added). Boi is looking to the ways in which general relativity and quantum field theory might be unified, arguing that the main work lies in the mathematics, not in experimental results. On such a view, expressed in a Husserlian idiom, the *essence* of matter-energy is fundamentally geometric, defined in the space-time geometry of the physical theory. Nature would then be "mathematized" in a very strong way along the lines that concerned Husserl in the *Crisis*.

I should like to recast the Friedmanesque account of physical theory within a Husserlian account of scientific theory. Briefly, the Husserlian model replaces the Kantian distinction between a priori and empirical or a posteriori parts of physical theory with a variation on the Bolzanoan distinction between purely logical and substantive or nonlogical parts of a theory. The point will not be that physical geometry is *epistemologically* prior to, or a priori relative to, physical theory, but something rather different. The whole theory, we may say with both Husserl and Quine, carries evidence that begins in perceptual observation. (Einstein's hypothesis of curved space was confirmed, after all, by observations of light bending as it passed the sun.) For Husserl, geometric theory carries a further kind of evidence from "eidetic" reflection, complicating the neat Quinean holistic picture of evidence. However, my concerns lie instead with the *ontological* structures posited in *phenomenological* meaning structures in the overall physical theory.

Husserlian Theory of Science with Restricted or Relativized Formal Ontology

In his later years Einstein told a friend, "The physicists say that I am a mathematician, and the mathematicians say that I am a physicist" (Overbye, 2004). Einstein's quip hits the nail right on the head from a Husserlian perspective: Physical theory is partly mathematical or formal, but only partly so. In Friedman's reconstruction, physical theory, whether Newtonian or Einsteinian, factors into a geometrical theory and a theory of gravitation and motion, the latter presupposing the former. Now Husserl's theory of theory distinguishes the levels of language, meaning, and objects represented. On the level of ontology, looking to the objects or objective structures represented in a theory, Husserl distinguishes formal and material ontology. I propose that *within the physical theory* we view the geometric theory as a restricted formal ontology and the theory of gravitation and motion as a material ontology presupposing that formal ontology. That is, *relative* or *restricted* to that physical theory, the geometric theory of the structure of space-time plays the role of a *formal* ontology that is presupposed by and structures the *material* ontology in the theory of gravitation and motion. (In the spirit of the quote from Boi, 2004, above, the mathematics of space-time is the formal ontology of the physical domain.)

When Husserl lays out his pure logic as the theory of theory in the *Logical Investigations*, he assumes logical distinctions of categories of expression and then emphasizes the correlation between categories of meaning and objective categories of things represented by meanings. (*Logical Investigations*, Prolegomena, § 66–68) Then he correlates "the theory of the possible forms of theory" with "the pure theory of manifolds" (§ 69–70). Broadly, Husserl thinks of the forms of theory along the lines of deductive axiomatic theories (like Hilbert). But then comes Husserl's innovation. Corresponding to the form of a theory, according to Husserl, is a manifold (Mannigfaltigkeit) of objects, taken as the "objective correlate" of a possible theory (§ 70). Husserl's notion here is an early form of the structuralist philosophy of mathematics, whereby a mathematical theory defines a purely formal structure that may be realized in, or applied to, different "domains" (see Smith, 2002a). In the present context, however, what is interesting is the place of geometry in Husserl's conception of our theory of nature.

Corresponding to formal categories of meaning, in Husserl's system, are "pure" or "formal" "objective categories" such as "Object, State of Affairs, Unity, Plurality, Number, Relation, Connection etc." (*Logical Investigations*, § 67). Formal categories—such as Object, Relation, and State of Affairs (correlates of the logical categories Name, Predicate, and Sentence)—apply invariantly to objects in any domain. By contrast, "material" categories apply to entities within a particular substantive or material domain such as Nature, Culture (Geist), or Consciousness (see *Ideas I*, and Smith, 1995). For Husserl, the theory of number is part of formal ontology, because numbers apply to objects in any possible domain, but the theory of space is part of the material ontology of nature. Arithmetic and geometry are then the mathematizations of our everyday conceptions or theories of number and space, respectively. Now Husserl sees an extension of the formal or mathematical in his conception of manifolds.

Mannigfaltigkeit, in the sense of multiplicity, was one of Cantor's terms for sets. But Husserl seeks a radical generalization of the notion of manifold, or Mannigfaltigkeit. Here Husserl looks explicitly to non-Euclidean geometry in Grassmann, Hamilton, Lie, Cantor, Riemann, and Helmholtz (*Logical Investigations*, § 70). In seeking a formal generalization of geometrical theory, he writes:

> If we use the term "space" of the familiar type of order of the world of phenomena, talk of "spaces" for which, e.g. the axiom of parallels does not hold, is naturally senseless. It is just as senseless to speak of differing geometries, when "geometry" names the science of the space of the world of phenomena. But if we mean by "space" the categorial form of world-space, and, correlatively, by "geometry" the categorial theoretic form of geometry in the ordinary sense, then space falls under a genus, which we can bound by laws, of pure, categorially determinate manifolds, in regard to which it is natural to speak of "space" in a more extended sense. Just so, geometric theory falls under a corresponding genus of theoretically interrelated theory-forms determined in purely categorial fashion, which in a correspondingly extended sense can be called "geometries" of these "spatial" manifolds. At any rate, the theory of *n*-dimensional spaces forms a theoretically closed piece of the

> theory of theory in the sense above defined. The theory of a Euclidian manifold of three dimensions is an ultimate ideal singular in this legally interconnected series of *a priori*, purely categorial theoretic forms (formal deductive systems). This manifold itself is related to "our" space, i.e. space in the ordinary sense, as its pure categorial form, the ideal genus of which the latter represents so as to say an individual singular rather than a specific difference. (*Logical Investigations*, § 70)

In *Ideas I*, we observed earlier, Husserl holds that the world is constituted through strata of meaning, beginning with the sense of time and then of space. If "our" world of nature is fundamentally structured by space and time, then we begin to see how the constitution of our world can begin with Euclidean space, plus a dimension of time, and move in refinement to a non-Euclidian space-time more accurate for astronomic scale. In our everyday perceptual experience, there is constituted time as we experience it and space as the frame of objects we see and move among in time; this is the "space" of everyday phenomena (things as experienced), and Euclidian geometry is a mathematical abstraction from our everyday understanding of this space. The notion of alternative geometries begins, however, with the ideal of a formal deductive system; by varying any axiom in the Euclidian system, we can then define alternative geometries. A system of geometry is then a formal deductive theory, and a geometrical theory is a system of propositions that constitute space in a certain way, that is, the theory defines or constitutes a certain manifold of spatial structure. That structure is the "formal" framework within which "natural" objects in that spatial or spatiotemporal framework occur and move.

Accordingly, we can reconfigure Friedman's account of physical theory in a Husserlian framework. Physical theory, Newtonian or Einsteinian, factors into two ranges of theory: a space-time geometry and a dynamics. The geometric theory plays the role of a formal ontology within the physical theory, as it applies to all objects in nature (all objects, whether individuals or events or states, occur in space or space-time). The dynamical theory presupposes the geometric theory and in turn plays the role of a material ontology within the physical theory, as it characterizes the laws of action and reaction (thus, we might say, the properly causal relationships) that hold between empirical bodies in space-time. That is to say, relative or restricted to the physical theory, the geometric and dynamic theories are respectively formal and material ontologies of nature. Thus at the level of language, tense logic defines expressions of time in what are *logical* terms in or relative to tense logic, and similarly, at the level of objective structure, geometric theory defines objective spatio-temporal structures in what are *formal* structures in or relative to physical theory.

Conclusion

In a Husserlian philosophy of science, scientific theories are tied into the surrounding world in several different ways. To begin with, a theory is a system of propositions (or the sentences that express them), and these propositions are themselves contents of (actual or potential) intentional acts of thought or judgment. From this perspective we distinguish three specific contextual features of science.

First, our scientific theories are evidentially grounded in everyday perceptions, and the *indexical* contents of these perceptions ("this . . .") prescribe objects in the immediate context of experience, that is, in our immediate surroundings within the natural and human world that is the object of scientific investigation. Second, our scientific theories typically involve *global*, or globally indexical, concepts, which appeal to essences of things in our surrounding world at large (as opposed to our immediate context of observation and theorizing), that is, our surroundings here on "this" planet or in "this" galaxy or in "this" universe. Third, both essences and concepts of things in nature, though ideal entities, are historically bound to the intentional activities of our scientific community's tradition of investigation, and, more specifically, to the ways in which these concepts were developed and the corresponding essences discovered.

Indeed, our physics remains tied into the historical situation of our experience in these three ways, even as our physical theory in general relativity and quantum mechanics appears to characterize essences of processes in nature that diverge from the essences posited in our everyday experience. The problem then becomes as follows: What is the relation between the essences posited in these realms of experience?

§ 2 The *Lebenswelt* in Husserl

Dagfinn Føllesdal

The idea of Husserl's that has become most widely known is no doubt that of the life-world. Particularly, the *word* "life-world" (*Lebenswelt*) has gained wide currency. The word was used by Simmel and others before Husserl (Simmel, 1912, 13).[1] After the Second World War, it became a favorite word of many social scientists, who used it in numerous different senses. Several of them refer to Husserl without seeming to have studied his philosophy and therefore without knowing the many important features that the life-world has in Husserl.

The first published work of Husserl's that contains the word Lebenswelt is his *Crisis*, of which the first two parts were published in 1936. The rest of the unfinished work, containing the important third part, with the main discussion of the life-world, was not published until 1954, but it was known to some of Husserl's students and followers, including Maurice Merleau-Ponty, who came to the Husserl Archives in Louvain to study this part in April 1939.[2]

Interpreters of Husserl differ widely in their views on the life-world. It is often thought that it constitutes a major break in Husserl's development, from the "early" Husserl of the *Ideas* to the late Husserl of the *Crisis*. Is it such a break? And secondly, what exactly is the life-world, and what role does it play in phenomenology? On the former question my answer is a definite "No." In what follows I will argue that the life-world is fully compatible with Husserl's earlier philosophy and that there is even a definite place for it in his phenomenology from its beginning. Husserl touches upon the life-world repeatedly in his earlier work, and he gradually deepens and modifies his views on it, as he did with everything else in his phenomenology. However, it is there all the time.

Rather than regarding the life-world as a break with Husserl's earlier philosophy, we should view the problematic of the life-world as intimately connected with the other main themes in phenomenology, particularly that of intentionality. To properly understand the life-world with all its nuances, it is important to fully appreciate the connection between it and the rest of Husserl's philosophy. In order to answer our second question, what is the life-world and what role does it play? I will start with a brief survey of the key ideas in Husserl's phenomenology and then show how the life-world is an integral part of his phenomenology from its inception.

Phenomenology as a Study of the Subjective Perspective

Phenomenology is an attempt to study in a systematic manner our different subjective perspectives, our different ways of experiencing reality. In the sciences one is searching for objectivity: One tries to secure that one's observations are independent of who makes them. They should preferably consist of reading numbers off a measuring scale or making other kinds of registrations that are affected as little as possible by the observer's subjective perspective. One does not deny the existence of a subjective perspective but regards it as a disturbing element when one is making scientific observations. One tries therefore to arrange experiments and observations in such a way that the influence of the subjective perspective becomes as small as possible.[3]

When our aim is to understand other persons, it does not suffice merely to know what impulses they are exposed to and how they move. It is far more important to know how they experience themselves and their surroundings. In order to understand what Husserl is after, we may begin with a simple example, the psychologist Jastrow's duck/rabbit picture, which was made famous by Wittgenstein. Actually, in order to come closer to Husserl, we will modify the example and consider not a picture but a silhouette of the real animal against the sky:

By permission of *Acta Philosophica Fennica.*

When we see this silhouette against the sky, we may see a duck or a rabbit. What reaches our eyes is the same in both cases, so the difference must be something coming from us. We structure what we see, and we can do so in different ways. The impulses that reach us from the outside are insufficient to uniquely determine which object we experience; something more gets added.

The Noema, Intentionality

This something more that gets added Husserl called the *noema*. The noema is a structure. Our consciousness structures what we experience. How it structures it depends on our previous experiences, the whole setting of our present experience, and a number of other factors. If we had grown up surrounded by ducks but had never even heard of rabbits, we would have been more likely to see a duck than a rabbit when confronted with the above silhouette; the idea of a rabbit would probably not even have occurred to us.

According to Husserl, our experience in a given situation can always in principle be structured in different ways; what reaches our senses is never sufficient to uniquely determine what we experience. Only in a few rare cases, such as in the duck/rabbit example, can we go back and forth at will between different ways of structuring our experience. Usually we are not even aware of any structuring going on; objects are simply experienced by us as having a structure.

The structuring always take place in such a way that the many different features of the object become connected with one another and are experienced as features of one and the same object. When, for example, we see a rabbit, we do not merely see a collection of colored patches, various shades of brown spread out over our field of vision. We see a rabbit, with a determinate shape and a determinate color, with the ability to eat, jump, and so forth. It has a side that is turned toward us and one that is turned away from us. We do not see the other side from where we are, but we see something that has an other side. Husserl expresses this also by saying that our consciousness is characterized by *intentionality*; it is always directed toward an object. That seeing is intentional, or object-directed, means just this: that the near side of the object we have in front of us is regarded as a side of a thing and that the thing we see has other sides and features that are co-intended, in the sense that the thing is regarded as more than just this one side. The noema is the comprehensive system of determinations that gives unity to this manifold of features and makes them aspects of one and the same object.

The word *object* must here be taken in a broad sense. It comprises not only physical things, but also, as we have just seen, animals and likewise persons, events, actions, and processes. When we experience a person, we do not experience a physical object, a body, and then infer that a person is there. We experience a full-fledged person; we are encountering somebody who structures the world, experiences it from his or her own perspective. Our noema is a noema of a person; no inference is involved. Seeing persons is no more mysterious than seeing physical objects; no inference is involved in either case. When we see a physical object, we do not see sense data or the like and then infer that there is a physical object there; our noema is the noema of a physical object. Similarly, when we see an action, what we see is a full-fledged action, not a bodily movement from which we infer that there is an action.

The noema is a key notion in Husserl's phenomenology, not only in his theory of perception, but in his analysis of all aspects of human consciousness. The noema is just Husserl's attempt to characterize the subjective perspective. Two people can face the same thing but nevertheless experience it quite differently, as differently as duck and rabbit. But even where they agree on what kind of an object they see, there may be enormous differences in the ways they see it. We all slide into routine ways of experiencing the surrounding world and ourselves. What characterizes a great artist is in part the ability to see and experience things, events, and persons in new ways, in part the ability to mediate this experience to others. Those objects that are being experienced need not be new and exceptional. They are often everyday and common. It is not the object but the experience that is central.

Filling

In the case of an act of perception, its noema can also be characterized as a very complex set of expectations or anticipations concerning what kind of experiences we will have when we move around the object and perceive it using our various senses. We anticipate different further experiences when we see a duck and when we see a rabbit. In the first case, we anticipate, for example, that we will feel feathers when we touch the object; in the latter case, we expect to find fur. When we get the experiences we anticipate, the corresponding component of the noema is said to be *filled*. In all perception there will be some filling: The components of the noema that correspond to what at present "meets the eye" are filled; similarly for the other senses.

Such anticipation and filling is what distinguishes perception from other modes of consciousness, for example, imagination or remembering. If we merely imagine things, our noema can be of anything whatsoever, an elephant or a locomotive standing here beside me. In perception, however, my sensory experiences are involved; the noema has to fit in with my sensory experiences. This eliminates a number of noemata that I could have had if I were just imagining. In my present situation, I cannot have a noema corresponding to the perception of an elephant.

This does not reduce the number of perceptual noemata I can have just now to one, for example, of you sitting there in front of me. It is a central point in Husserl's phenomenology that I can have a variety of different perceptual noemata that are compatible with the present impingements upon my sensory surfaces. In the duck/rabbit case, this was obvious; we could go back and forth at will between having the noema of a duck and having the noema of a rabbit. In most cases, however, we are not aware of this possibility. Only when something untoward happens, when I encounter "recalcitrant" experience that does not fit in with the anticipations in my noema, do I start seeing a different object from the one I thought I saw earlier. My noema "explodes," to use Husserl's phrase, and I come to have a noema quite different from the previous one, with new anticipations. This is always possible, says Husserl. Perception always involves anticipations that go beyond what at present meets the eye, and there is always a risk that we may go wrong, regardless of how confident and certain we might feel. Misperception is always possible.

Constitution

Objects are intended as having a great number of properties. Normally, as in the case of a material object, many more than can ever be exhausted in our experience of it. Objects are *constituted* through our consciousness, Husserl says. This does not mean that we create them or bring them about, merely that the various components of the noema are interconnected in such a way that we have an experience as of one full-fledged object. All there is to the object hence corresponds to components of our noema. In the case of physical objects, the inexhaustible character of what is experienced is a characteristic anticipation in our noema and hence an important feature of what it is to be a physical object. According to Husserl,

> An object "constitutes" itself—"whether or not it is actual"—in certain concatenations of consciousness which in themselves bear a discernible unity in so far as they, by virtue of their essence, carry with themselves the consciousness of an identical X. (*Ideas I*, § 135)

Incidentally, Husserl's use, here and in many other places, of the reflexive form "an object constitutes itself," is an indication that he did not regard the object as being produced by consciousness. Husserl considered phenomenology as the first strictly scientific version of transcendental idealism, but he also held that phenomenology transcends the traditional idealism-realism distinction, and in 1934 he wrote in a letter to Abbé Baudin: "No ordinary 'realist' has ever been as realistic and concrete as I, the phenomenological 'idealist' (a word which by the way I no longer use)."[4] In the Preface to the first English edition of the *Ideas I*, Husserl stated:

> Phenomenological idealism does not deny the factual [*wirklich*] existence of the real [*real*] world (and in the first instance nature) as if it deemed it an illusion . . . Its only task and accomplishment is to clarify the sense [*Sinn*] of this world, just that sense in which we all regard it as really existing and as

> really valid. That the world exists . . . is quite indubitable. Another matter is to understand this indubitability which is the basis for life and science and clarify the basis for its claim.[5]

The World, the Past, and Values

We constitute not only the different properties of things but also the relation of a thing to other objects. If, for example, I see a tree, the tree is conceived of as something that is in front of me, as perhaps situated among other trees, as seen by other people than myself, and so forth. It is also conceived of as something that has a history: It was there before I saw it, it will remain after I have left, perhaps it will eventually be cut and transported to some other place. However, like all material things, it does not simply disappear from the world.

My consciousness of the tree is in this way also a consciousness of the world in space and time in which the tree is located. My consciousness constitutes the tree, but at the same time it constitutes the world in which the tree and I are living. If my further experience makes me give up the belief that I have a tree in front of me because, for example, I do not find a tree-like far side or because some of my other expectations prove false, this affects not only my conception of what there is but also my conception of what has been and what will be. Thus in this case, not just the present but also the past and the future are reconstituted by me. To illustrate how changes in my present perception lead me to reconstitute not just the present, but also the past, Husserl uses an example of a ball, which I initially take to be red all over and spherical. As it turns, I discover that it is green on the other side and has a dent:

> The sense of the perception is not only changed in the momentary new stretch of perception; the noematic modification streams back in the form of a retroactive cancellation in the retentional sphere and modifies the production of sense stemming from earlier phases of the perception. The earlier apperception, which was attuned to the harmonious development of the "red and uniformly round," is implicitly "reinterpreted" to "green on one side and dented." (*Experience and Judgment*, § 21a)

So far I have mentioned only the factual properties of things. But their *value* properties are constituted in a corresponding manner, Husserl says. The world within which we live is experienced as a world in which certain things and actions have a positive value, others a negative. Our norms and values, too, are subject to change. Changes in our views on matters of fact are often accompanied by changes in our evaluations.

Horizon

When we are experiencing an object, our consciousness is focused on this object; the rest of the world and its various objects are there in the background as something we "believe in" but are not at present paying attention to. The same holds for most of the inexhaustibly many features of the object itself. All these further features of the object, together with the world in which it is set, make up what Husserl calls the *horizon* of that experience. The various features of the object, which are co-intended, or also meant, but not at the focus of our attention, Husserl calls the *inner horizon*, while the realm of other objects and the world to which they all belong he calls the *outer horizon*:

> Thus every experience of a particular thing has its *internal horizon*, and by "horizon" is meant here the *induction* which belongs essentially to every experience and is inseparable from it, being in the experience itself. The term "induction" is useful because it suggests [*vordeutet*] (itself an "induction") induction in the ordinary sense of a mode of inference and also because it implies that the latter, for its elucidation to be completely intelligible, must refer back to the original, basic anticipation.
>
> However, this aiming-beyond [*Hinausmeinen*] is not only the anticipation of determinations which, insofar as they pertain to this object of experience, are now expected; in another respect it is also an aiming-beyond the thing itself . . . to other objects of which we are aware at the same time, although at first they are merely in the background. This means that everything given in experience has not only an internal horizon but also an infinite, open, *external horizon of objects cogiven*. . . . These are objects toward which I am not now actually turned but toward which I can turn at any time. . . . all real things which at any given time are anticipated together or cogiven only in the background as an external horizon are known as real objects (or properties, relations, etc.) *from* the world, are known as existing within the one spatiotemporal horizon. (*Experience and Judgment*, § 8)

Take a simple example of an item belonging to this outer horizon. If I had asked you as you were entering this room what your expectations were, you might mention something about friends you expected to meet, a lecture you expected to hear, and so on. It is highly unlikely that you would mention that you expected there to be a floor in the room. Yet as I saw you confidently stepping in, I have every reason to believe that you expected that there would be a floor here. You did not think about it—your attention was directed to other things—but you had a disposition that you acted on, and also, if I had asked you whether you expected there to be a floor in the room, you might have wondered why I asked such a trivial question, but you would probably have answered yes.

Expectations and beliefs are dispositional notions. We count as beliefs not only thoughts that we are actively entertaining but also those that we rarely think about, for example, that $2 + 2 = 4$. We do have a problem when we try to delimit exactly what beliefs we have. The method of questioning is not reliable. On the one side, it gives too much. Remember how in the *Menon* a skilled questioner uncovers that the slave boy has the most unexpected geometrical beliefs. Plato took this as evidence for his theory of anamnesis. On the other, it yields too little. As Freud and others have taught us, we often sincerely deny that we have beliefs that seem all too apparently to underlie our actions.

The most reliable criterion, which we often fall back on, is to assume that people have those beliefs that best explain their actions, including their verbal activities. However, then a further problem is that the states we appeal to in order to explain people's actions are not exclusively cognitive states. Also various physical states are needed, and skills of various kinds that it is often hard to classify as mental. Thus while our arithmetical skills are presumably mental, our skills in swimming or walking can hardly be classified as mental. Then we have tricky intermediate cases, such as one's keeping a standard distance to partners in a conversation, where the standard may vary from culture to culture. Is keeping this distance a matter of a tacit belief that it is the proper distance? Or is it a matter of a bodily skill that is gradually acquired as one grows up in this culture? And what about the way we sign our name?[6] Obviously, cognitive activities are involved in the process that brought us to sign it the way we do: We had to learn the alphabet, we had to learn our name, and so on. But also, in our semiautomatic way of signing it, bodily skills are involved to a great extent. Various personality traits play a role, as do certain general traits of our culture.

Opinions concerning examples such as these may vary. One of our problems is that we have no clear-cut way to settle such issues, lacking a precise definition of what is to count as mental, what as physical. However, there is obviously an interplay here, both in the process that leads to the skill and in the skill itself, and any satisfactory theory of intentionality must heed such an interplay. The noema may still be defined as a structure, but the anticipations that are related through this structure are not merely the anticipations involved in seeing, hearing, and the like, but also the anticipations involved in kinesthesis and bodily movement, where when something "goes wrong," we are aware of it. We are familiar with this experience of "going wrong" from cases of misperception: We cannot always tell exactly what went wrong, but we are aware that something went wrong.

As I tried to show in my contribution to the *Festschrift* for Hintikka's fiftieth birthday (Føllesdal, 1979), Husserl started out in the *Ideas* and other early works with a strongly cognitivist attitude: The anticipations in the noema are of a purely cognitive kind. However, in manuscripts from 1917 on, he focused more and more on the role of the practical and the body in our constitution of the world. He never worked out the full implications of this for his conception of the noema, but it seems clear that he would consider our anticipations as not merely beliefs but also as bodily settings, which are involved in kinesthesis and also play an important role in perception and in the movements of our body. In numerous passages, some of which I quoted in that article, Husserl talks about practical anticipations and the role of kinesthesis in perception and bodily activity.

The Noema, the Horizon, and the World

It is sometimes thought that the noema comprises only some of the anticipations that we have at a given time, those that correspond to the object that we are attending to and its properties. All the other anticipations, which concern the rest of the world, are then not included in the noema. Husserl never explicitly states his position on this point. However, there is no basis in Husserl's writings for such a restricted view of the noema. On the contrary, many passages point in the direction of identifying the noema with the *total* set of anticipations, including those that relate to the outer horizon and the world. Thus when he introduces the noema in the *Ideas* he says:

> The transcendent *world* receives its "parenthesis" . . . With the whole physical and psychical *world*, the actual existence of the real relation between perceiving and perceived is parenthesized, and, nonetheless, a relation between perceiving and perceived (as well as between liking and liked) remains left over. (*Ideas I*, § 88)

Also when Husserl wants to emphasize that the noema is not changed by the bracketing, he includes in that which is bracketed not just the isolated individual thing that we perceive but the whole world: "The full noema . . . is not touched by excluding the actuality of the tree and that of the whole world" (*Ideas I*, § 97).

Further, Husserl seems to hold that the whole horizon is encompassed by the noema; thus, for example, he writes in a manuscript:

> The horizon-structure with its levels and penetrations is the noematic framework of sense and validity through which objectivity is in each case meant for us. (*Manuscript B II 9*, p. 146)

While Husserl is not explicit concerning the terminological issue of how many of our anticipations the noema is supposed to comprise, all of them or only some, he leaves no doubt concerning the systematic point that is our main concern. In all his writings from the *Ideas* on, he is quite clear that *when we intend an object, we co-intend at the same time the whole world to which that object belongs.*

The Iceberg

The horizon is of crucial importance for Husserl's concept of justification, which we shall discuss later. What is particularly significant is the hidden nature of the horizon. As we noted, the horizon is that which is not attended to. Usually, as in the case of the floor in this room example above, we have not even thought about it. Typically, we cannot even recall when we first acquired the corresponding belief or anticipation. According to Husserl, there may never have been any occasion when we actually judged there to be a floor in some particular room. Still we have come to anticipate a floor, not in the sense of consciously expecting one, but in the sense that if we entered the room and there were none, we would be astonished. In this example we would easily be able to tell what was missing; in other cases our anticipations are so imperceptible that we just may feel that something has gone awry, but we may not be able to tell what it is.

Words like *belief* and *anticipate* are clearly not the proper ones here, since they have overtones of something being conscious and thematized. Both English and German seem to lack words for what we want to get at here: Husserl uses the words *antizipieren* and *hinausmeinen* and also *vorzeichnen*.

Life-World and Natural World

The problematic of the life-world is intimately connected with the distinction between the natural attitude and the transcendental, or phenomenological, attitude, which Husserl introduced in 1906–07. The first mention of the problematic for which he later introduced the term *Lebenswelt* occurs shortly thereafter, in his lectures "Grundprobleme der Phänomenologie" in 1910–11, that is, already before the *Ideas*. Husserl begins these lectures with an extended discussion of "Die natürliche Einstellung und der 'natürliche Weltbegriff,'" that is, "The Natural Attitude and the 'Natural World-Concept.'" He here says:

> It could also be shown that philosophical interests of the highest dignity require a complete and comprehensive description of the so-called *natural world concept*, that of the natural attitude, on the other hand also that an accurate and profound description of this kind is not easily disposed of, but on the contrary would require exceptionally difficult reflections. (*Husserliana XIII*, 124.34–125.7)

Husserl here borrows the phrase *natürlicher Weltbegriff*, natural attitude, which he emphasizes, from Avenarius, whom he discusses later in the lecture (Avenarius, 1891). In a manuscript from 1915, Husserl describes, following Avenarius, this world in the following way:

> All opinions, justified or unjustified, popular, superstitious, scientific, all relate to the already *pregiven* world. . . . All theory relates to this immediate givenness and can have a legitimate sense only when it forms thoughts which do not offend against the general sense of the immediately given. No theorizing may offend against this sense. (*Husserliana XIII*, 196.22–34)

In the following years, Husserl repeatedly returns to this and related themes, using various labels that sometimes allude to other philosophers who had propounded similar ideas, such as Nietzsche. Quite often he uses Avenarius's phrase *natürliche Welt*, natural world. In a manuscript from 1917, which appears to be the first place he uses the word *Lebenswelt*, he introduces this new word as equivalent to the former: "The life-world is the natural world—in the attitude of the natural pursuit of life are we living functioning subjects involved in the circle of other functioning subjects" (*Husserliana IV*, 375.31–33).[7]

Gradually through the 1920s and especially in the 1930s, the life-world becomes a central theme in Husserl's writings, until his discussion culminates in the *Crisis* in 1936. A main point of this latest of Husserl's works was to open a new and better access to phenomenology, through the notion of the life-world. The life-world is for Husserl our natural world, the world we live in and are absorbed by in our everyday activities. A main aim of phenomenology is to make us reflect upon this world, make us see how it is constituted by us. Through a special reflection, the "phenomenological reduction," phenomenology will take us out of our natural attitude where we are absorbed by the world around us into the phenomenological, transcendental attitude, where we focus on the noemata of our acts, on our structuring of reality. That is, we perform the "parenthesizing," or "bracketing," that was described by Husserl in the first passage from the *Ideas* that was quoted in the section above on the noema, the horizon, and the world.

Pregivenness and Intersubjectivity

In the passage I just quoted from Husserl's 1915 manuscript, Husserl says that the world is pregiven (*vorgegeben*). This point is also discussed in the *Ideas*, where Husserl notes:

> In my waking consciousness I find myself in this manner at all times, and without ever being able to alter the fact, in relation to the world which remains one and the same, though changing with respect to the composition of its contents. It is continually "on hand" for me and I myself am a member of it. Moreover, this world is there for me not only as a world of mere things, but also with the same immediacy as a world of values, a world of goods, a practical world. (*Ideas I*, § 27)[8]

And a few pages later, he writes:

> I continually find at hand as something confronting me a spatiotemporal reality [*Wirklichkeit*] to which I belong like all other human beings who are to be found in it and who are related to it as I am. (*Ideas I*, § 30)[9]

In these same sections of the *Ideas I*, Husserl stresses the shared, intersubjective nature of this world. He does this particularly in § 29, which he titles "The 'Other' Ego-Subjects and the Intersubjective Natural Surrounding World." He there says:

> I take their surrounding world and mine Objectively as one and the same world of which we are conscious, only in different modes [*Weise*] . . . For all that, we come to an understanding with our fellow human beings and together with them posit an Objective spatiotemporal actuality. (*Ideas I*, § 29)

The same ideas of pregivenness and intersubjectivity are repeated with almost the same words when Husserl discusses the life-world in the *Crisis*, for example in § 37, where he says:

> The life-world, for us who wakingly live in it, is always there, existing in advance for us, the "ground" of all praxis, whether theoretical or extratheoretical. The world is pregiven to us, the waking, always somehow practically interested subjects, not occasionally but always and necessarily as the universal field of all actual and possible praxis, as horizon. To live is always to live-in-certainty-of-the-world.

In the *Crisis*, as in the *Ideas*, Husserl stresses intersubjectivity; thus, for example, in § 47:

> Thus in general the world exists not only for isolated men but for the community of men; and this is due to the fact that even what is straightforwardly perceptual is communal.

Was the Life-World a Late Development in Husserl?

As I mentioned in my introduction, it is often thought that the life-world constitutes a major break in Husserl's development, from the early Husserl of the *Ideas* to the late Husserl of the *Crisis*. One of the foremost exponents of this view is David Carr, who in *Interpreting Husserl* writes:

> It may be thought that the concept of the life-world really represents nothing new in Husserl's thought . . . . But as Husserl's exposition unfolds it begins to take on features that distinguish it in more than just emphasis from what has gone before.
>
> One such feature is the prominence of the notion of *Vorgegebenheit* or pregivenness. . . . A *second* feature which distinguishes the account of the perceived world in the *Crisis* from earlier accounts, is that this world is repeatedly described as public or intersubjective: it is "pre-given as existing for all in common," as the "world common to us all." (1987, 232–34)

The passages I have quoted from Husserl's early writings and manuscripts make it amply clear that there is considerable continuity in early and late Husserl concerning the life-world and also concerning the two features of the life-world mentioned by Carr. Rather than stressing the differences, I would like to stress the continuity in Husserl's development. In particular, I find it helpful when we try to understand Husserl to note the intimate connection between his notions of noema, horizon, and life-world.

One Life-World or Many?

However, we are now in for a problem: Husserl seems to contradict himself with regard to the number of life-worlds. From what has been said so far, we should expect that we all have different life-worlds, depending on our cultural background, our past experiences, and so on. Husserl also says this, for example in his lectures on *Phänomenologische Psychologie* from 1925:

> We do not share the same life-world with all people, not all people "in the world" have in common with us all objects which make up our life-world and which determine our personal activity and striving, even when they come into actual association with us, as they always can (to the extent that, if they are not present, we come to them and they to us). (*Husserliana IX*, 496.40–45)

This also fits in well with Husserl's repeated contention that as our opinions change, so does our life-world; thus, as we shall see later, the development of science changes our life-world. On the other hand, Husserl also insists that there is just *one* life-world:

> The world, on the other hand, does not exist as an entity, as an object, but exists with such uniqueness that *the plural makes no sense* when applied to it. Every plural, and every singular drawn from it, presupposes the world-horizon. (*Crisis*, § 37, my emphasis)

We may, of course, conclude that Husserl here contradicts himself, and we may opt for one of the two alternatives as the one that best represents his view: either a plurality of life-worlds or just one. However, when we interpret Husserl sympathetically, there is no contradiction, and we can accept both views. They both express important features of the life-world. To read Husserl this way, we only have to keep in mind the general recurring theme in all of phenomenology, namely, that one and the same object can appear in many different ways. Let us first see in what sense there is just one life-world:

There is *one* life-world in the following sense: There is one world that appears to a person in each of his or her many experiences, and this same world is also the world that is experienced by everybody else, regardless of when and where they might live. Our conceptions of this world may differ, and in this sense we all live in *different* life-worlds. But different is not distinct. The world in which each of us lives is one and the same, but it appears differently to each of us.

That the world appears differently to each of us goes without saying. But the oneness of the life-world is an important point for Husserl, which deserves our attention. The idea that there is only one world, in which we all live, underlies intersubjectivity and communication. Both intersubjectivity and communication are premised on the idea of there being just one world, which we experience from our different perspectives. Another highly important feature of the *one* life-world is touched on in the last sentence in the last quotation from Husserl above: "Every plural, and every singular drawn from it, presupposes the world-horizon." This theme is a very important one in Husserl's discussion of the life-world, and it is clarified further many places in his works. Most illuminating is § 37 of the *Crisis*, which I quoted earlier, in connection with the pregivenness of the life-world.

For Husserl, when we take a stand with respect to the existence or non-existence of an object, the truth or falsity of a statement, when we value something positively or negatively, and have goals and purposes for our lives, this happens always within the life-world, and, what is important, it happens only *in virtue of* the life-world. As Husserl expresses it, the life-world is the founding of validity (*Geltungsfundierung*) or the soil (*Boden*) for this taking of a stand.[10] It is "already there," and it serves as an "*unhintergehbar*" background (one which one cannot get around) for all our taking of stands. For example, we can always affirm or deny the existence of an object, but we cannot deny the existence of the world. We may try to do so, but our words have no "*behauptende Kraft*," or affirmative force, to talk with Frege. We shall come back to this in the final section of this paper, where we shall discuss the role of the life-world in justification.

Science and the Life-World

A contested point in Husserl scholarship is the relation between the life-world and the sciences. Many interpreters of Husserl like to find in the life-world an opposition to the sciences. However, such a disdain for the sciences is out of character with Husserl's background in and continued interest in mathematics and science. It also accords poorly with the texts, which give a different and more intriguing picture. According to Husserl, the life-world and the sciences are intimately connected, in three different ways:

1. The world of science is *part* of the life-world. This comes out most explicitly and clearly in *Experience and Judgment*, where Husserl says:

> Everything which contemporary natural science has furnished as determinations of what exists also belong to us, to the world, as this world is pregiven to the adults of our time. And even if we are not personally interested in natural science, and even if we know nothing of its results, still, what exists is pregiven to us in advance as determined in such a way that we at least grasp it as being in principle scientifically determinable. (*Experience and Judgment*, § 10)

Similar statements are found also in other of Husserl's writings, for example in the *Crisis*: "Now the scientific world—[the subject matter of] scientific theory— . . . like all the worlds of ends 'belongs' to the life-world" (*Crisis, Appendix VII*, p. 380).

2. Scientific statements get their *meaning* by being embedded in the life-world. This was stressed by Husserl already in the manuscript from 1915 that I quoted above in the section on the life-world and the natural world:

> All opinions, justified or unjustified, popular, superstitious, scientific, all relate to the already pregiven world. . . . All theory relates to this immediate givenness and can have a legitimate *sense* only when it forms thoughts which do not offend against the general sense of the immediately given. No theorizing may offend against this sense. (*Husserliana XIII*, 196.22–34)

3. The sciences are *justified* through the life-world. There is an interplay between this point and point one above: The sciences are justified because they belong to the life-world, and at the same time, they belong to the life-world because they are conceived of as describing the world, as claiming to be true:

> Though the peculiar accomplishment of our modern objective science may still not be understood, nothing changes the fact that it is a validity for the life-world, arising out of particular activities, and that it belongs itself to the concreteness of the life-world. (*Crisis*, § 34f)

And similarly:

> All these theoretical results have the character of validities for the life-world, adding themselves as such to its own composition and belonging to it even before that as a horizon of possible accomplishments for developing science. The concrete life-world, then, is the grounding soil [*der gründende Boden*] of the "scientifically true" world and at the same time encompasses it in its own universal concreteness. (*Crisis*, § 34e)

Ultimate Justification

This brings us to the final theme of this paper: the role of the life-world in *justification*. In an earlier paper I have argued—against the traditional "foundationalist" Husserl interpretation— that Husserl both in science and in ethics had a view on justification similar to that of Goodman and Rawls (Føllesdal 1988, 107–29). An opinion is justified by being brought into "reflective equilibrium" with the *doxa* of our life-world. This holds even for mathematics: "Mathematical evidence has its source of meaning and of legitimacy in the evidence of the life-world" (*Crisis*, § 36).[11]

A major puzzle that many see in this idea of justification is "How can appeal to the subjective-relative doxa provide any kind of justification for anything? It may help to resolve disagreements, but how can it serve as justification?" Husserl answers by pointing out that there is no other way of justifying anything, and that this way is satisfactory:

> What is actually first is the "merely subjective-relative" intuition of prescientific world-life. For us, to be sure, this "merely" has, as an old inheritance, the disdainful coloring of the *doxa*. In prescientific life itself, of course, it has nothing of this; there it is a realm of good verification and, based upon this, of well-verified predicative cognitions and of truths which are just as secure as is necessary for the practical projects of life that determine their sense. The disdain with which everything "merely subjective and relative" is treated by those scientists who pursue the modern ideal of objectivity changes nothing of its own manner of being, just as it does not change the fact that the scientist himself must be satisfied with this realm whenever he has recourse, as he unavoidably must have recourse, to it. (*Crisis*, § 34a)

So far, this is a mere claim. However, Husserl elaborates his view in other parts of his work. His key observation, which I regard as an intriguing contribution to our contemporary discussion of ultimate justification, is that the "beliefs," "expectations," or "acceptances" that we ultimately fall back on are unthematized, and in most cases, have never been thematized. Every claim to validity and truth rests upon this "iceberg" of unthematized prejudgmental acceptances that we discussed earlier. One should think that this would make things even worse. Not only do we fall back on something that is uncertain, but on something that we have not even thought about and have therefore never subjected to conscious testing. Husserl argues, however, that it is just the unthematized nature of the life-world that makes it the ultimate ground of justification. Acceptance and belief are not attitudes that we decide to have through any act of judicative decision. What we accept, and the phenomenon of acceptance itself, are integral to our life-world, and there is no way of starting from scratch, or "to evade the issue here through a preoccupation with aporia and argumentation nourished by Kant or Hegel, Aristotle or Thomas" (*Crisis*, § 34e). Only the life-world can be an ultimate court of appeal:

> Thus alone can that ultimate understanding of the world be attained, behind which, since it is ultimate, there is nothing more that can be sensefully inquired for, nothing more to understand. (*Formal and Transcendental Logic*, § 96b)

§ 3 The Origin and Significance of Husserl's Notion of the *Lebenswelt*

Ulrich Majer

If it is a sign of the philosophical genius that he creates and establishes *new* notions, concepts, and ideas in the traditional philosophy of his time, then Husserl was, without doubt, a great philosopher: During his whole career as philosopher, he introduced many new notions, beginning with the verb "*kolligieren*" in his first book *The Philosophy of Arithmetic* and ending with the concept of the *Lebenswelt*, or life-world, in *The Crisis of European Sciences and Transcendental Phenomenology*, his last major work, first published in 1936 in the journal *Philosophia*. But things are often not as simple as they seem at first glance. The notion of the Lebenswelt is simple and difficult at the same time; it depends on how one understands it—as a *commonsense* notion or as a *philosophical* concept or as *both*, as Husserl seems to do; he is at least in this respect quite consistent.

In this chapter I will investigate the origin and significance of Husserl's notion of the Lebenswelt. I will not, however, begin with this delicate task before having clarified one point of possible misunderstanding. As many readers likely know, the book version of *The Crisis of European Sciences* emerged from a lecture that Husserl had presented in Vienna and Prague in 1935 (*The Vienna Lecture*), and which was later published in the Husserliana edition. Toward the end of this piece, he gives a short summary of his *diagnosis* of the crisis and its possible *solutions* from which I will quote some central remarks.

> The crisis of European Existence, talked about so much today and documented in innumerable symptoms of the breakdown of life, is not an obscure fate, an impenetrable destiny; rather it becomes understandable and

> transparent against the background of the *teleology of European history* that can be discovered philosophically. The condition for this understanding, however, is that the phenomenon Europe be grasped in its central, essential nucleus. In order to be able to comprehend the disarray of the present crisis, we had to work out the *concept of Europe as the historical teleology of the infinite goals of reason*; we had to show how the European world was born out of ideas of reason, i.e., out of the spirit of philosophy. The crisis could then become distinguishable as the *apparent failure of rationalism*. The reason for the failure of a rational culture, however, as we said, lies not in the essence of rationalism itself but solely in its being rendered superficial, in its entanglement with naturalism and objectivism.
>
> There are only two escapes from the crisis of European existence: the downfall of Europe in its estrangement from its own rational sense of life, its fall into hostility toward the spirit and into barbarity; or the rebirth of Europe from the spirit of philosophy through a heroism of reason that overcomes naturalism once and for all. (*The Vienna Lecture*, p. 299)

Now I must confess that the *antiscientific* attitude underlying Husserl's talk and his diagnosis of the crisis and its desired cure by overcoming the *naturalism* and *objectivism* of the natural sciences (including mathematics) is, from my point of view, highly problematic, if not simply mistaken. But this does *not* imply that something very interesting and highly valuable could not be found in Husserl's late work. On the contrary, I am convinced that Husserl's invention of the concept of the Lebenswelt as an "irreducible" ingredient of all sciences is a very important discovery—not only from a philosophical but also from a scientific point of view. Consequently, it is my main task (1) to separate what seems to me right and what seems to me wrong in Husserl's diagnosis of the crisis and (2) to explain in what sense Husserl's analysis of the situation should be taken seriously. But the question immediately arises: By *whom* should it be taken seriously? By the philosopher, by the scientist, or by both? Here we encounter a first difficulty, and it is by no means the only one. There are similar difficulties regarding the notions of naturalism and objectivism. Do they denote a *philosophical* position, or do they refer to certain aspects of science that Husserl believes to lie at the root of its crisis?

In spite of these difficulties, and in light of Husserl's notoriously cryptic and long-winded style of writing, it seems to me advisable to analyze Husserl's concept of the Lebenswelt in a somewhat unusual manner. Instead of presenting an immanent interpretation of his late writings, I will approach them from an "external" point of view. I will compare what Husserl has to say on *The Crisis of European Sciences* with the utterances of some prominent scientists who have dealt —implicitly or explicitly— with similar questions and problems regarding the methodological foundations as well as the epistemological limits of science. I have in mind above all Weyl and Hilbert, who both stood in more or less close scientific contact with Husserl during different periods of their lives. After a brief review of Husserl's considerations with respect to the indispensable role of the Lebenswelt in science, I will turn first to Weyl and discuss his conception of "Science as a Symbolic Construction of Man," the title of an important essay in which Weyl criticizes a certain view of science as "ridiculously circular." Next, I will turn to Hilbert and explain the sense in which he regards the Lebenswelt as an irreducible presupposition of all sciences, geometry and mathematics included. (Although Hilbert does not use the term Lebenswelt, he speaks of the geometry of "everyday life" and similar expressions.) Finally, I return to Husserl and his diagnosis of the crisis and try to separate what seems to me a justified critique of the positivistic worldview of science and what has to be rejected as misguided or even absurd. The essential result will be that the natural sciences and their *foundations* are not fixed and finished but are still open to supplementation and fundamental revision in the future.

A Brief Review of Husserl's *Crisis*

The ambiguity in the title of this section is quite deliberate. In fact, it has been noticed by some authors that Husserl's book *The Crisis of European Sciences* is itself the result of a kind of crisis in Husserl's own thinking. It has been argued, for instance by Lothar Eley, that Husserl in the *Crisis* became aware that he had left unsolved a fundamental problem in his very first book *The Philosophy of Arithmetic*, namely the role and legitimation of the actual *infinite* in his genetic approach to arithmetic. Let me quote from Eley's Introduction to that book:

> To the true philosophy of the calculus also belong investigations of his late work *The Crisis of European Sciences*. This late work refers back in a remarkable manner to his early work [*The Philosophy of Arithmetic*] as we will see in a moment. According to this [genetic] program, the paradise of the actual infinite would have been dispelled from mathematics. Therefore, it comes as

> a surprise to rediscover this paradise in the late work *The Crisis of European Sciences*. (*Philosophie der Arithmetik, Husserliana XII*, p. xiii)

I will not now decide whether Eley is correct, whether Husserl really solved the old problem of the role of the actual infinite in mathematics in his late work. Instead I want to point out that the title of *The Crisis of European Sciences and Transcendental Phenomenology* contains a word that seems rather strange and out of place in this context. Why does Husserl not simply say "The Crisis of Sciences" instead of speaking explicitly of "The Crisis of *European* Sciences"? There must be a reason for this, and, in fact, the reason is not difficult to come by: It is connected with the circumstance that the sciences, as we know them today (mathematics included), had their first origin in Greece and their rebirth during the Renaissance in Italy, after many centuries of deep sleep, and from there spread over Europe and the rest of the world. Now this story is such a *commonplace* that it alone cannot be the point of Husserl's considerations. And indeed, if we look more closely at Husserl's text, we recognize another strand. It is the cultural history of "European mankind," of which the natural sciences form only a part, that is at the focus of Husserl's concerns. Unfortunately the term *cultural history* is still somewhat too broad to characterize precisely what Husserl had in mind. For it is not, of course, the cultural history of European mankind as seen by a normal historian or a sociologist or one or another scientist, but, unsurprisingly, as seen by a *philosopher*. He, the philosopher, has a special interest in the contributions of philosophers to the cultural history of mankind, and this means the cultural history of European mankind, because it was in Europe that philosophy became the leader in the cultural evolution of mankind. This (and the next point, to which I will come in a moment) is much clearer in the short lecture held in Vienna than in the long book.

> In this lecture I shall venture the attempt to find new interest in the frequently treated themes of the European crisis by developing the philosophical-historical idea (or the teleological sense) of European humanity. As I exhibit, in the process, the essential function that philosophy and its branches, our sciences, have to exercise within that sense, the European crisis will receive a new elucidation. (*The Vienna Lecture*, p. 269)

But what is the "function" or *role* of philosophy and its branches in this historical sense, and what is the European crisis of which Husserl speaks in the beginning and which later—after the elucidation—turns out to be the crisis of European *sciences*? In order to make a very long story short, I will condense the answer into six propositions:

1. Philosophy has been the *leader* in the intellectual evolution of man—not only in its very beginning in Greece two and a half thousand years ago, but also in its further development, in particular during the Renaissance.

2. The main reason for the success of this intellectual evolution was the circumstance that philosophical considerations gave birth first to mathematics and then to the natural sciences.

3. The first indications of a crisis had occurred already when mathematics and the sciences began to emancipate themselves from philosophy by developing their own standards of objectivity and truth. This was in the beginning a very slow and almost unnoticeable process, which only lately became accelerated by the successes of science.

4. The crisis broke out (and became public for the first time) when the sciences—in spite of their successes—*divorced* themselves from philosophy and imposed, in turn, their own canonical standards of "objectivity" and "truth" not only on psychology—and all other disciplines of mind and culture—but also on philosophy.

5. This does not imply, however, that philosophy has lost its leadership and authority in the intellectual evolution of man once and for all. It can regain its leading role and authority if it can make clear that the sciences, including mathematics, are not so completely sovereign as they themselves believe, but rest on presuppositions that they cannot escape.

6. The totality of these presuppositions is later called the Lebenswelt, and it is the task of *philosophy*—not the sciences—to investigate the content and meaning of the concept of Lebenswelt and to demonstrate the relevance and significance of this concept for the sciences.

Although these statements are my own formulations, it should be clear that they do not express my own opinion. I wanted only to capture Husserl's convictions as outlined in the Vienna lecture and later elaborated at great length in *The Crisis of European Sciences*. From now on, I will keep a more *critical* distance from Husserl's point of view. My first critical point is the following:

My sketch of Husserl's *Crisis* (and even more, of course, the book itself) can give the impression that Husserl had written a history of philosophy enriched by some very interesting subchapters on the history of modern science. But this, I think, is a mistake. The history that Husserl tells in the first chapters is much too sloppy and, what is more important, far too *permeated* with common prejudices and trivial errors for it to count as a serious attempt to write a book on the history of philosophy, not to mention of science—at least it fails to meet my standards in these matters. I would not only demand more historical precision but also expect a deeper and more concrete understanding of what was going on intellectually in a given historical development. Instead, I think, Husserl's intention was a completely different one: He wanted to put certain philosophical ideas (regarding his theory of mind and intentions) into a "historical dressing" so that it seemed to the reader as if he had only to strip the ideas of their historical clothing in order to get a clear and genuine view of them.

Now this already sounds more critical than I wanted, but I have, at least for the first part of my thesis, some good corroboration in the Vienna lecture. Not only is the entire line of this lecture more theoretical than genuinely historical, but Husserl also makes clear that history has an only instrumental character for him.

> The foregoing considerations on the philosophy of the spirit provide us with the proper attitude for grasping and dealing with our subject of spiritual Europe as a problem purely within the humanistic disciplines [*ein rein geisteswissenschaftliches Problem*], and first of all in the manner of spiritual history. As already indicated in our introductory statements, a remarkable teleology, inborn, as it were, only in our Europe, will become visible in this way, one which is quite intimately involved with the outbreak or irruption of philosophy and its branches, the sciences, in the ancient Greek spirit. We can foresee that this will involve a clarification of the deepest reasons for the origin of the portentous naturalism, or, what will prove to be equivalent, of modern dualism in the interpretation of the world. Finally this will bring to light the actual sense of the crisis of European humanity. (*The Vienna Lecture*, p. 273)

Now the passage just quoted cries out for clarification. What does "purely humanistic" (*rein geisteswissenschaftlich*) mean? Does Husserl intend a "science of mind" in distinction and opposition to the natural sciences? If the answer is yes, as I suspect, then it is highly problematic, because it sets up a wrong opposition. What does "naturalism" mean, and why is it fatal? And why is the so-called modern dualism *equivalent* to naturalism and therefore equally fatal? These are all legitimate questions, but I will not go into them now. I only want to show that Husserl is not doing "real" history of philosophy or of science, as the *historian* understands this task, but instead tries to advance an idea of mind, which he projects backward into the history of mankind, notably European mankind! One can find such claims on almost every other page of the Vienna lecture. Here are two more examples:

> "The spiritual shape of Europe"—what is it? [We must] exhibit the philosophical idea which is immanent in the history of Europe (spiritual Europe) or, what is the same, the teleology which is immanent in it, which makes itself known, from the standpoint of universal mankind as such, as the breakthrough and the developmental beginning of a new human epoch—the epoch of mankind which now seeks to live, and only can live, in the free shaping of its existence, its historical life, through ideas of reason, through infinite tasks. (*The Vienna Lecture*, p. 274)

> Universal philosophy, together with all the special sciences, makes up only a partial manifestation of European culture. Inherent in the sense of my whole presentation, however, is that this part is the functioning brain, so to speak, on whose normal function the genuine, healthy European spiritual life depends. The humanity of higher human nature or reason requires, then, a genuine philosophy. (*The Vienna Lecture*, pp. 290–91)

I think the last quotation makes explicit that Husserl's intention is *not* to investigate history as we find it in written documents, but to *model* history so that it fits nicely with his idea of a new type of mankind, a mankind whose highest goal is *reason*, where the approach to reason is understood as an *infinite* task. To be aware of this circumstance is quite important for my further considerations for two practical reasons.

First, it relieves me of the difficult and boring duty of checking whether Husserl's historical claims and suggestions are *historically* correct. I can simply take them as mediate expressions of his philosophical intentions and opinions dressed in historical settings; whether they are correct and concordant with the historical documents is another matter. This does not mean that it is unimportant to know where differences exist; on the contrary, such differences tell a lot about Husserl's *true* philosophical intentions and opinions; this is in particular relevant with respect to the most elaborated historical parts of the *Crisis*, in which Husserl presents his view of the genesis of modern science as a "mathematization of nature" by Galileo. Now this view is not only a platitude, but it is also rather misleading, because the real innovation was to my mind something quite different, namely the fact that Galileo and his successors invented genuine physical *experiments* and made quantitative predictions and measurements. I will come back to this point.

Second, knowing that the highest goal of mankind is *reason* in the sense of an *infinite* task makes it somewhat easier to grasp what the systematic message, the philosophical point of Husserl's excursions into the history of science and philosophy actually is. We have only to take two further steps: First, we must make clear to ourselves what the supposed "crisis of the sciences" consists in. Second, and more important, we must understand what the philosophical significance, the epistemological essence of the pretended crisis is. Once we are clear on these two points, we will see why the crisis—or at least its *ghost*, in the case that it does not really exist—can only be banished, according to Husserl, if we go back to the so-called Lebenswelt and make a fresh start in understanding our own existential situation in the world.

So much, for the time being, on Husserl's *Crisis*. Of course, I have not yet explained what the core of the crisis is, let alone its philosophical significance. But, as I said in the beginning, I will not give an immanent interpretation but instead offer an external explication by comparing Husserl's *Crisis* with similar considerations in the writings of Hilbert and Weyl. Before I start my detour, let me make two brief remarks, the first concerning a philological point, the second a conceptual one.

The expression Lebenswelt does not occur literally in the Vienna lecture; but the *concept* as such is already present, as the following quotation from the lecture shows. Having made a distinction between two attitudes or habits (the theoretical attitude of the Greek philosophers and the more practical attitudes in other cultures), Husserl remarks with respect to the relation of these two attitudes:

> Thus the theoretical attitude, in its newness, refers back to a previous attitude, one which was earlier the norm; [with reference to this] it is characterized as a reorientation. Universally considering the historicity of human

> existence in all its communal forms and in its historical stages, we now see that a certain attitude is essentially and in itself the first, i.e., that a certain norm-style of human existence (speaking in formal generality) signifies a first [type of] historicity within which particular factual norm-styles of culture-creating existence remain formally the same in spite of all rising, falling, or stagnating. We speak in this connection of the natural primordial attitude, of the attitude of original natural life, of the first originally natural form of cultures, whether higher or lower, whether developing uninhibitedly or stagnating. All other attittudes are accordingly related back to this natural attitude as reorientations [of it]. (*The Vienna Lecture*, pp. 280–81)

One could continue this quotation at length, but the passage in question suffices to show that the term Lebenswelt roughly denotes what Husserl describes here as the original attitude or habit of the "original natural life" (*des ursprünglich natürlichen Lebens*). This is important for my exegesis because it makes clear that Husserl could have chosen another expression, for example Hilbert's expression "everyday life" (*tägliches Leben*) instead of Lebenswelt, which indeed, I think, influenced his choice.

The second clarification relates to the notion of reason as an *infinite task*. What does Husserl mean by this? A relatively short answer is given when Husserl considers the first decisive *Umstellung*, or change of attitude, in the history of mankind, the step of the Greek philosophers from the attitude of natural life toward a *theoretical* habit:

> The historical course of development is prefigured in a determined way by this attitude toward the surrounding world. Even the most fleeting glance at the corporeity to be found in the surrounding world shows that nature is a homogeneous, totally interrelated whole, a world by itself, so to speak, encompassed by homogeneous space-time, divided into particular things, all being alike as *res extensae* and determining one another causally. Quite rapidly, a first and great step of discovery is taken, namely, the overcoming of the finitude in spite of its open endlessness. Infinity is discovered, first in the form of idealized magnitudes, of measures, of numbers, figures, straight lines, poles, surfaces etc. Nature, space, time, become extendable *idealiter* to infinity and divisible *idealiter* to infinity. From the art of surveying comes geometry, from the art of numbers arithmetic, from everyday mechanics mathematical mechanics, etc. (*The Vienna Lecture*, p. 293)

This is a very important section, not only because in it Husserl indicates what the "infinite task of reason" is, but also because it makes clear that he distinguishes between the *finiteness* of nature, which is no more than an open *endlessness*, and the *infinite* (in mathematics) as an *idealization* of certain concepts and their underlying intuitions. We find a similar distinction in Hilbert's writings between the *finite* but *unlimited* nature and the *infinite* in mathematics as ideal elements in our theoretical conception of nature. This suggests that Hilbert and other philosophically minded scientists like Weyl, Schrödinger, Bohr, and Heisenberg (to name only the best-known quantum theorists) may have encountered problems similar to those Husserl describes in the *Crisis.* So we should perhaps consider what they have to say about the same or at least some closely related issues. But here already I encounter a first difficulty: *Which* issue?

Until now I have not said very much about the *proper issue* of Husserl's late work, *The Crisis of European Sciences.* But one thing seems quite obvious: The scientists themselves do not speak of a "crisis of the sciences"—at least not in the 1930s. So why should we assume that they treat the same or at least a similar problem as Husserl in the *Crisis*? There seems to be no reason that they should. But contrary to this initial impression, the scientists in question did think about the *methods* and *limits* of their own science repeatedly; and this alone seems to me reason enough to ask whether they came across problems similar to Husserl's, even if they expressed them differently and came to quite different conclusions. A first problem, which it seems to me has a certain resemblance to Husserl's crisis of the sciences (or at least to a certain aspect of it), is Weyl's "ridiculous circle."

Weyl's "Ridiculous Circle"

Toward the end of his 1949 essay "Wissenschaft als symbolische Konstruktion des Menschen" (Science as Symbolic Construction of Man), Weyl touches some open questions and problems regarding the foundations of quantum mechanics. His main concern is the problem of measurement in quantum mechanics and its different treatments in Copenhagen and Zürich, namely Bohr's philosophy of *complementarity* on the one hand and Schrödinger's appeal to a kind of *dialectic* in the recognition of nature on the other. Dealing with the latter, its sense and justification, Weyl detects a certain resemblance between Schrödinger's *dialectic* of recognition and a position that he addresses as the *Existentialphilosophie.* In this context Weyl develops the ridiculous circle I mentioned previously as an argument in defense of Schrödinger's dialectic, I suppose, and hence of the so-called existential philosophy.

Before I investigate what the latter has to do with Husserl's *Crisis*, let me briefly explain the ridiculous circle. As scientists, we may be tempted, says Weyl, to argue in the following way:

> As we know, the chalk on the blackboard consists of molecules, and these are composed of charged and uncharged elementary particles, electrons, neutrons, etc. However, in analyzing what we mean with such words in theoretical physics, we have seen that the physical things get dissolved into a *symbolism* manipulated according to certain rules; the symbols themselves, however, turn out in the end to be concrete signs written with chalk on the blackboard. You recognize the *ridiculous circle* [*den lächerlichen Zirkel*]. We can escape from this circle only if we accept the manner in which we understand in daily life man and things, which we encounter, as an irreducible fundament. (Weyl, 1968, 342, my italics, my translation)

Now, I can very well imagine that most readers will agree that this circle is really ridiculous, but also that many readers will, at the same time, be very skeptical about two points: First, does anyone exist who directly and seriously affirms such a ridiculous circle or is *guilty* of affirming it indirectly because he labors under such a delusion? Second, what does the circle, what do Weyl's considerations regarding the problem of measurement have to do with Husserl's *Crisis*? Let me first answer the second, easier, question. Here I am on relatively safe ground, because some remarks in Husserl indicate that both authors have the same point in mind, or at least a very similar one. Considering the relation of the natural sciences to *our*, that is, to the human mind, Husserl remarks:

> Is it not absurd and circular to want to explain the historical event "natural science" in a natural-scientific way, to explain it by bringing in natural science and its natural laws which, as spiritual accomplishment, themselves belong to the problem? (*The Vienna Lecture*, p. 273)

Surely, this is not exactly the same circle as Weyl's ridiculous one, but it is closely related to the latter. This becomes obvious once we take the broader context of Weyl's circle argument into account. I will do this in several steps. First, we must recall that the title of Weyl's essay is "Science as *Symbolic* Construction of Man," and this reveals that symbols, the symbols of science, are first and foremost *mental* products, products of the human mind (*mind* taken as a general term under which many individuals fall). So far Husserl and Weyl agree. That the symbols have to be written down on stone or paper or in semiconductors in order to store them so that we may communicate by means of them over space and time, is—seen from this perspective—only a contingent matter. This becomes clear if we consider the context in which Weyl discusses the circle argument more carefully. For this is not just the problem of measurement in quantum physics, as one might guess from what I have said so far, but also the foundations of *mathematics*, and in particular—and this may come as a surprise—Hilbert's so-called formalism.

> Here now I detect certain relations between quantum physics and the foundations of mathematics, and *existential philosophy*. When Bertrand Russell and others tried to reduce mathematics to pure logic, a *residue of meaning* [*Bedeutung*] remained in the form of the simple logical concepts. But in Hilbert's formalism this residue disappears as well. Nonetheless we need signs, real signs written with chalk on the blackboard or with ink on paper. We have to understand what it means to place one stroke beside the other; it would be wrong to reduce this naively and roughly understood spatial order of signs to a *purified* spatial intuition and structure as it is expressed for example in Euclidean geometry. Instead we have to rely on the natural understanding in dealing with things in our natural world surrounding us. (Weyl, 1968, 341)

I have quoted this paragraph at full length, because it is extremely important. It contains the key to a proper understanding of Husserl's concept of the Lebenswelt. First of all, one has to understand that Weyl accepts Hilbert's formalism not only as *tolerable* but also as *inevitable* for a proper reflective understanding of mathematics. Because that has not always been the case, it is much more remarkable that Weyl had changed his mind in this respect already in 1926. Second, one must grasp that, although signs are necessary in order to have concrete objects of intuition, very little, indeed almost nothing, depends on their physical garb in Hilbert's approach to mathematics. Whether you choose dots or strokes, chalk or ink (or something else) does not matter as long as one can *recognize, distinguish,* and *order* them. Third, and this is the crucial point, it would be in vain to develop a *scientific theory of symbols* by offering an exact geometrical, physical, or chemical analysis of *signs*. This would be a *category mistake* of the worst kind, comparable only with a confusion of body and mind, to which the confusion of sign and symbol bears a certain resemblance. For a *sign* to become the representative of a *symbol*, it suffices that the sign be a simple concrete recognizable object of our common intuition. To require a geometrical or physical analysis of a *sign* qua *symbol* is a confusion of a symbol qua an *object of mind* with a sign as a *physical mark* on paper or stone, which leads inevitably into the circle. Although both are bound closely together, symbol and sign are *not identical.*

What I have just pointed out with respect to the connection of sign and symbol and their different *epistemological* role and *ontological* status is, of course, not only valid in pure mathematics but also in the natural sciences, that is, in physics for the relation between the theoretical concepts (symbols) and the measurement of their values by means of real rulers and compasses (signs) in concrete experiments. The relation, as Weyl takes it, is quite analogous:

> Just as we manipulate [*umgehen*] concrete signs in Hilbert's formalized mathematics, so in physics, when we perform measurements and their necessary operations, we manipulate boards, wires, screws, cog-wheels, pointer and scale. We move here on the same level of understanding and action as the cabinet-maker or the mechanic in his workshop. (Weyl, 1968, 342)

I hope you see now the intimate relationship between Weyl's ridiculous circle and Husserl's remark regarding the *absurd* attempt to *explain* the sciences as an historically grown product of our mind by means of the sciences themselves. Furthermore, I hope you recognize why in both cases only a *recursion* to the so-called Lebenswelt can protect us against the danger of becoming a victim of the ridiculous circle. So far so good. But what is the so-called Lebenswelt? What does Husserl mean by this expression? How should we understand this term? A first answer, I think, can be extracted from Weyl's last remark: "*Lebenswelt*" means a mode of life, a mode of human existence in which *no theoretical knowledge* is required, but only some *practical abilities* of understanding and acting are supposed, like those of the craftsman. Only in this way, only by recurring to this mode of life, can we avoid the ridiculous circle.

Now let me come back to the first question: Does there really exist somebody, maybe a scientist or a philosopher, who affirms something like Weyl's ridiculous circle? At first glance, this seems *improbable*, because the circularity of the circle is so obvious (at least in Weyl's presentation) that nobody will be guilty of it. But on second thought, doubts creep in: Is there not a tendency to explain everything *naturalistically* by the laws of physics and the other natural sciences, leaving nothing unexplained? Are there not scientists, such as Watson and Crick, who are proud that they have found no trace of the human mind or consciousness in their laboratories and quickly conclude that these things can only *exist*, if they exist at all, in the form of *extremely complex physical structures* (like the double helix)? From here it is only one step to philosophical positions like *reductionism* or strong *physicalism*, which look to me more like a blank check on the future than the result of serious research.

Now in spite of such doubts, the affirmation of the question of the *real existence* of the ridiculous circle gets further support: Couldn't it be the case that Husserl had positions like the one just mentioned in mind, when he spoke of the "crisis of European science," and was it not Weyl's intention to brand certain opinions held by some scientists as *absurd* by his example of a ridiculous circle? The answer seems to be *yes* in both cases, but before I articulate this more specifically, I have to clarify two points in order not to be misunderstood.

1. What I have said so far looks as if I supposed that Husserl and Weyl subscribe to a kind of mind-body dualism. But this is definitely not my intention, nor does it correspond to my own convictions. I wish to be neutral in this respect. I only use the mind-body terminology as a comfortable mode of speech to refer to different domains of well-distinguished phenomena. Whether these phenomena can be "reduced" onto one another or not, I will leave open for now. My intention to be neutral does not, however, imply that I will not criticize *naive* forms of physicalism or reductionism. On the contrary, I share the view that an absolute, unconditional physicalism, without any limits, is in danger of very quickly becoming absurd. I argue for more patience in these matters.

2. The question of whether the danger of a ridiculous circle is real or only a product of the imagination of some philosophers is difficult to answer, because it is infected by an *ambiguity* that I cannot completely resolve. The ambiguity arises because scientists do not form a homogeneous group. Some are rather naive positivists, who are convinced that it is only a question of time until it will be shown that

> we are *nothing but* computers or aggregates of molecules or whatever you like; others are much more careful and reflected. Hence, the question arises: To whom shall we refer if we want to answer the question of whether the circle or the crisis is real or only a bogeyman of philosophers? My strategy will be the following:

First, I will argue that serious scientists neither assert nor fall victim to something like Weyl's ridiculous circle; quite the contrary, they warn against such short circuits. Second, I will suggest that there is a problem in science regarding the intricate relation of body and mind (taken in the broadest sense), which is not only unsolved, but in respect to which scientists are at present relatively helpless and, consequently, postpone its solution to the future. This invites philosophers and other visionaries, who have insufficient patience, to extend the methods and results of physics beyond the limits of inanimate nature to the very different domain of living beings and their social, cultural and, last but not least, *mental* achievements. It seems to me, and this is my third and last point, that this is the *right* place for Husserl to step in with his analysis and, of course, with his critique of the crisis of European sciences. If I am not mistaken, it is the point at which he does in fact intervene.

So let's come to the serious scientists first and inquire which stance they take with respect to the circle and the related problems of the *Crisis*. We know already that Weyl is sympathetic to Husserl's rejection of scientism. But there are, of course, many more serious scientists—so many that I cannot name them all, let alone present their views in this matter. Therefore I proceed differently: I choose the champion of scientific optimism, who maintained that *every* problem in science that can be stated clearly can also be solved. This is, of course, no one other than Hilbert. He is most liable, because of his scientific optimism, to become a victim of the ridiculous circle. Consequently, if I can show that this is not the case, that Hilbert is quite aware of the circle, we have a good argument for supposing that the other serious scientists also resist the danger of a short-circuit.

I have to restrict my analysis of Hilbert's scientific work to three points. First and most important: Hilbert acknowledges—as do Husserl and Weyl—the existence of something like the Lebenswelt (he refers to it as "the domain of everyday life") as a *precondition* of science. Second, in two public lectures held in Copenhagen and Zürich (in 1921 and 1923, respectively),[1] he makes a sharp distinction between "inanimate nature" as the proper object of his world-equations (and other physical theories) and the domain of the life sciences, to which we humans with our conscious actions and cultural institutions belong. Last but not least, Hilbert denies that it is possible to expand the range of physics in such a way that it becomes a *theory of everything*, including our mind and our thinking, because this leads inevitably into paradoxes, in fact a whole bundle of paradoxes.

Unfortunately, I do not have the space to go into detail on all these points. I have done this elsewhere for the first and second.[2] With respect to the third point, let me briefly explain what is going on: In the lecture *Natur und mathematisches Erkennen* (1919) and again in the lecture *Über die Einheit in der Naturerkenntnis* (1924), Hilbert tackles the question of whether it is possible to complete and finish physics—*vollenden* and *abschließen* are the German words—in the sense that the resulting world-model encompasses literally everything. His answer to this question is a definite no, because this would lead to a number of paradoxes, of which the following is for our purposes the most interesting.

> Finally, the spiritual domain as well, our thinking in particular, would have to be something merely apparent [*bloß Scheinbares*]—an absurd consequence for a view of nature which arises from the desire to make all content of reality accessible to our thinking.

Summarizing his reflections, Hilbert comes to the following conclusion:

> Through this [and the other] paradoxes we are forced to conclude that the assumption of the *perfectability* [*Vollendbarkeit*] of physical knowledge in an all-embracing theory is not admissible, and hence, that the ideal of a perfectly completed physical knowledge remains inaccessible in principle. (Hilbert, 1924, 180, my emphasis)

Now I think this conclusion would meet with the full agreement of Husserl, because what Husserl tried to capture and to *condemn* by his notion of naturalism is precisely this: It is not permitted on pain of absurdity to extend the natural sciences and their "*Weltbild*" (world-picture) to an all-embracing theory, capturing everything, including our own thinking and theorizing about nature. So far so good. But the interesting point in the present context is now this: If this is correct, Husserl's own position is incoherent, to say the least, because he cannot speak of a crisis of the *sciences* if there are scientists like Hilbert who quite consciously warn against a "totalitarian" conception of nature, such as a complete and finished theory of the sort just mentioned, a "theory of everything" so to speak, which should, for the sake of "self-consistency," be avoided under all circumstances.

Of course, one might think that Hilbert is only the usual exception to a general rule. Most scientists would not be so careful and reflective as Hilbert, who as a mathematician and logician of first rank has a very good nose for lurking paradoxes. This may well be, but my impression is a different one. Looking through the writings of quite a number of the great physicists, one finds in most cases similar reflections and critical remarks like Hilbert's. I have already mentioned Weyl, but this is also true of Schrödinger, Bohr, and Born, and also of Heisenberg. If this is correct (and I see no reason why anyone should doubt this), then Husserl is at least forced to make a *distinction* between different types of scientists: those who reflect on their own activities as natural scientists and recognize and accept certain limits of science and those who don't, succumbing to the temptation of an uncritical scientism, or naturalism, as Husserl calls it. But then the crisis is not a crisis of the sciences themselves but of a certain brand of uncritical scientists (who indeed should be watched suspiciously and criticized, if necessary). To this I can agree without any reservation. But I suspect Husserl's proper message is a different one: He wants to scrutinize and criticize his own discipline, that is, philosophy, its history, and future development. But this is another topic that must be consigned to further research. Let me instead close with a remark about quantum theory and its relation to Husserl's notion of Lebenswelt.

If there ever has been the danger that science itself would become a victim of Weyl's ridiculous circle, then this danger existed—and still exists to a certain degree—in quantum physics. The reason is the following: Because quantum theory is one of the fundamental theories of all physics, the temptation is quite real to describe the measuring process in terms of quantum mechanics and to treat the whole system, that is, the quantum object and the measuring apparatus, as one complex interacting object of quantum theory itself. But very soon it turned out that this was not feasible, neither practically nor theoretically, and that a distinction had to be made, called *cut*, which divided the complex object again into a proper object of quantum theory and a measuring device. The interesting point with respect to Husserl is that the measuring apparatus had to be both treated and described as a device that does not belong to the "proper" objects of quantum mechanics—it had to be described somehow "classically." This is Bohr's famous principle of complementarity. It could also be called Husserl's principle of "the indispensability of the Lebenswelt" for a proper noncircular understanding of science itself. How quantum theory and Lebenswelt are interrelated is a difficult question, which presumably will take a long time before it can be answered perspicuously without getting entangled in absurdities. I do not, however, believe, as Husserl seems to do, that a philosophical analysis of the term Lebenswelt will help very much in this respect. I presume instead that an investigation of the question "What is life?" and hence, a corresponding sedimentation of the meaning of the term *Leben,* will be the better approach.

§ 4 Husserl on the Origins of Geometry

Ian Hacking

Husserl aux pantoufles[1]

Ursprung (*Origin*) is a wonderfully *enthusiastic* piece of writing. I like to imagine it as welling up in Husserl all his life, as something he desperately wanted to say. It has the excitement of adolescence about it more than the precision of the mature scholar. Or rather the nostalgia of an old man in carpet slippers recalling the delights of youth, a bit like the second movement of Mahler's second symphony. That is called the *Resurrection* symphony. After the first movement buries the young hero, the second is all recollected lyricism and love tunes, before we carry on with the serious business of resurrection in the remaining three movements. Mahler sometimes but only sometimes called the second movement irrelevant. One might sometimes but only sometimes also call *Ursprung* irrelevant. It demands careful reflection but not carping criticism. I hope that my remarks below, although in a style and tradition entirely different from Husserl's, can partake of Husserl's sense of sheer joy and wonder. Naive astonishment should set the tone for all serious discussion.

On Husserl's Awe in the Face of Geometry

"What Mathematics Has Done to Some and Only Some Philosophers": Under that title I argued that some philosophers have been overwhelmed by the phenomenon of mathematical demonstration (Hacking, 2000, 83–138). Plato, Leibniz, Kant, Bertrand Russell, Ludwig Wittgenstein, and Imre Lakatos figure prominently in my discussion. Each, in his way, allowed a concern for mathematical proof to infect his whole philosophy. Other philosophers are notable for their lack of interest, and perhaps they are right to ignore the phenomenon of mathematical knowledge, even to disdain it. But I have always suspected that some and only some people are bowled over by proofs, while others simply fail to grasp their power. Proof brings with it the idea of a priori knowledge and necessary truth.

The feeling of astonishment, even of awe, that comes with understanding a notably deep and perspicuous demonstration is one of the things that has moved the philosophers. In my opinion, it is a mark of profound philosophical sensibility to be so moved and distinguishes the important thinkers from lesser minds. My list of philosophers struck by mathematical proof did not include Husserl, but he was among the awed.

I shall be discussing what mathematics did to Husserl. I shall not claim that it invaded his whole philosophy, but it certainly moved him deeply from the time of his early interest in the foundations of arithmetic to his 1935 lecture published in 1939 as "Die Frage nach dem Ursprung der Geometrie als intentionalhistorisches Problem," which henceforth I shall call *Ursprung.* And it certainly did infect his thinking about the exact sciences in the *Crisis.* He focused not on proof but on geometry as the science of ideal objects. But as I shall explain, you cannot have ideal objects unless you also have the practice of making and telling proofs. It is proofs that ensure—or create?—the objectivity of mathematics that was so much the concern of the *Crisis.* By proof, of course I do not mean Leibnizian, combinatorial, line-by-line proofs, but what I have already called perspicuous proofs, proofs that one can in the Cartesian way take in whole at once, by what Husserl would have called a direct intuition of its validity.

Bigger Game

I share Husserl's awe in the face of geometry. Perhaps like Plato, he is more impressed by idealized shapes, while I am more impressed by perspicuous proofs. But Husserl was after far more than the origins of geometry. Even the second half of the title that was retrospectively given to his lecture—"as intentional-historical problem"—does not get at the half of it. The inquiry is to be "exemplary." It aims at carrying out, "in the form of historical meditations, self-reflections about our own present philosophical situation in the hope that in this way we can finally take possession of the meaning, method, and beginning of philosophy, the *one* philosophy to which our life seeks to be and ought to be devoted" (*Origin of Geometry*, p. 354). I do not share this project, which is not to say that I dismiss it, only that I cannot participate in it.

In addition, *Ursprung* was written at the same time as the *Crisis* and was probably intended to be included in it. That is a book written by a German nationalist proud to claim intellectual descent from the grand line of German philosophers, indeed the "European" philosophers, the tradition that begins in Greece and is fully transmitted only to Germany. Husserl was dismayed by the irrationalism that was sweeping his nation. He was wholly German, with a son who had died for the Fatherland at Verdun. But he was also an elderly Jew (he had retired from his chair in 1928). His best pupil had overtaken his own role as the most prominent German philosopher—with a philosophy that, in his judgement, was another symptom of the pervasive irrationalism of the time. The abiding theme of *Crisis* was to be a return to (what Husserl thought was) the greatest European contribution to world history, namely the discovery of rationality and objectivity.

Looking back, as we do now, the book is a work of tragedy. And that in a strong sense. It was doomed to fail in its attempt to restore the state of reason. First, because any purely intellectual endeavor would, no matter how impassioned, be impotent against the raging politics of violence and hate. Worse, that politics had its rival vision of the greatness that was Europe. Second, the book was doomed because its analysis of the deep ground of pending disaster was just wrong. Whatever be the merits of phenomenology, it was not the case that Europe fell into the abyss of self-destruction because European science had lost touch with something deep and original in the human spirit, a depth that underlay the sciences and made them possible. I shall not further discuss these matters, but it is important to recall that *Ursprung* aims at more than an analysis of the origins of geometry.

Primal Beginnings

Over and over we read in the essay words that are translated as *primal sources, the primal beginnings* (for example, *Origin of Geometry*, p. 367). These are the *Ur* words. When it is beginnings in time, Husserl tended to use the word *Urstiftung*. The title word *Ursprung* means not simply origin but source. In the case of knowledge, it is not only the source of the knowledge but also the ground, the evidence, for that knowledge. I have called the present chapter "Husserl on the Origins [plural] of Geometry." It would be misleading to call it "Husserl on the Origin [singular] of Geometry." Yes, this is a chapter about Husserl's *Ursprung*, the lecture, but I have no commitment to the notion that geometry has a singular origin, or to the idea of a primal evidence.

I distrust the lust for things primal. There is, for sure, an ineluctable drive in Western consciousness to find the first moment. To find the skeleton of the first human being. To scan the first three seconds of the universe. To reveal the primal scene, or to relive the primal scream. Each of those programs makes sense, although some may prove to be wrongheaded or illusory.

I do believe that geometry had beginnings. Undoubtedly, there were several beginnings. In these days of postcolonial history, one notes that Husserl lived in the confident world where civilization mostly meant European civilization and its Mediterranean predecessors. The history of ancient mathematics had been codified in the mid-nineteenth century by mostly German and a few French and British scholars. It was a cumulative story that led from Babylonia via Athens and Alexandria to Göttingen. Today the texts of that original history are being reread with new eyes. Other civilizations are being opened up. Karine Chemla and Shuchun Guo have just published the most classic of Chinese texts, *The Nine Chapters*. To quote from Geoffrey Lloyd's preface:

> One can thus say that the classical mathematics of ancient China, and Euclid's *Elements*, represent two radically different styles of mathematical reasoning. This comparison teaches a crucial lesson: that there is no unique pattern of development that mathematics has to follow. (Chemla and Guo, 2004, xi)

Lloyd goes on to say that the existence of different styles of mathematical reasoning shows that there was no one thing at which mathematicians were aiming and that it is important for Westerners to grasp the existence of alternative styles.

Perhaps Lloyd took the so-called axiomatic method of Euclid as definitive of the Greek style. Husserl states that axioms are not what is fundamental in Greek mathematical reasoning. I agree, although perhaps for a reason different from Husserl's. In my opinion the making of proofs was fundamental.

The ability to make and grasp proofs must be a human universal, although a few people in any culture are much better at it than others. It is utterly immaterial whether the discovery of this human faculty occurred only once, among some peoples on the eastern shores of the Mediterranean, or many times in many places. My problem is that there is nothing metaphysically deep to learn about this discovery, nothing primal, nothing that leads to a depth problem or a depth solution bearing on that philosophy to which our life ought to be devoted.

Perhaps all quests for the primal will turn out to be a mistake. The idea of the primal scene has been trashed, most effectively not by the Freud bashers but by the Wolfman himself. The primal scream has echoes, now, only in the murmurings of psychobabble. You may think this reference is flippant, but no. Husserl's idea of the primal, of the Ur, is curiously close to that of Freud's primal scene. Both, of course, have roots in Hegel. Freud's thought was that if only the Wolfman could fully recover the scene of his parents' copulation in certain positions—and the reworking of it after latency—then he would understand and overcome the neuroses that swamped him later in life. Now whatever might be the imagined mechanism by which the scene worked on the Wolfman's unconscious, there was no doubt that it was supposed to be *deep*, that is, of profound significance. I find it interesting but not deep, and have exactly the same respectful skepticism about Husserl's primal sources.

Kant's Awe

The classic statement of naive astonishment at geometry is to be found in one of the passionate passages to be found in Kant. I suppose everyone knows it, but please allow me to quote the entire paragraph in translation. I shall repeat certain words in square brackets as a sort of running tally of the points being made.

> In the earliest times of which history affords us a record, *mathematics* had already entered on the sure course of science, among that wonderful nation, the Greeks [*the Greeks*]. Still it is not to be supposed that it was as easy for this science to strike into, or rather to construct for itself, that royal road, as it was for logic, in which reason has only to deal with itself [*easy for logic,*

hard for maths]. On the contrary, I believe that it must have remained long—chiefly among the Egyptians—in the stage of blind groping after its true aims and destination, and that it was revolutionized by the happy idea of one man, who struck out and determined for all time the path which this science must follow, and which admits of an infinite advancement [*revolution, one man*]. The history of this intellectual revolution—much more important than the discovery of the passage round the celebrated Cape of Good Hope—and of its author has not been preserved [*a revolution whose importance can hardly be overestimated*]. But Diogenes Laertius, in naming the supposed discoverer of some of the simplest elements of geometrical demonstration—elements which, according to ordinary opinion, do not even require to be proved—makes it apparent that the change introduced by the first indication of this new path, must have seemed of the utmost importance to the mathematicians of that age, and it has thus been obscured against the chance of oblivion. A new light must have flashed on the mind of the first man (*Thales*, or whatever may have been his name) who demonstrated the properties of the *isosceles* triangle. [*A new light flashes on a single mind!*] For he found that it was not sufficient to meditate on the figure, as it lay before his eyes, or the conception of it, as it existed in his mind, and thus endeavour to get at the knowledge of its properties, but that it was necessary to produce these properties, as it were, by a positive *a priori construction* [*a proof, by constructing these properties from that concept*]; and that in order to arrive with certainty at an *a priori* cognition, he must not attribute to the object any other properties than those which necessarily followed from that which he had himself, in accordance with his conception, placed in the object. [*It is not just observing the properties of the triangle, but demonstrating it by a construction—one that is a priori and needs no measurement.*] (Kant, 1924, Bxi–xii; translated by Meiklejohn)

Husserl agreed with the content of Kant's prehistory of geometry and shared in its spirit. This does not go without saying, and that for several reasons. First, although he comes back to the legend of Thales, Husserl is evidently thinking of communal activity: "the first oral cooperation of the beginning geometers" (*Origin of Geometry*, p. 368). The emphasis on community is an essential supplement to Kant. But there is a beginning all right. "Clearly, then, geometry must have arisen out of a *first* acquisition, out of first creative activities" (*Origin of Geometry*, p. 355). The recent disciplines known as science studies or sociology of scientific knowledge insist that all science is social. Husserl realized that long ago, while Kant seems not to have done so.

There is a second consideration. Husserl is only imperfectly aligned with Kant. Kant emphasizes proof, a construction out of the concept that yet does not go beyond the concept. Husserl accentuates the ideal objects of geometry, the ideal objects that fall under the idealized concept. But in any "principle-of-charity" reading of the two philosophers, the difference must be superficial. For only by demonstrating properties of the object (say Kant's isosceles triangle) is there any sense of the ideal object, an object that of necessity has these properties. Some readers of Husserl would insist that Husserl wants us to form a direct acquaintance with the ideal object, but that is to miss what Husserl saw: It is perspicuous proof that allows us to directly intuit the objects.

Conversely, if there is certainty about the properties, apodictic certainty (to use Kant's word *apodictic*, which Husserl makes his own), then we are not speaking of the empirical isosceles, but of an ideal object.

Galileo's Mathematization of Nature

After the long passage just quoted, Kant turns to seventeenth-century physics. It matters little to the sequel, but we should record a fundamental difference between Kant's attitude and that of Husserl. Kant does say that he is confining himself "to the *empirical* side of natural science," yet it is clear that he believes that "the wise Bacon gave a new direction to physical studies." Galileo figures because he "experimented with balls of a definite weight on the inclined plane." After the achievements of Galileo and Torricelli and Stahl, "a light broke upon all natural philosophers." Kant's hero was Newton. Kant's Galileo introduces the "experimental method." He added in a footnote that "the first steps" of that method "are involved in some obscurity." Indeed the contributions of Galileo and Stahl are separated by about a century!

Husserl, finishing his book in 1935, never mentioned Bacon. The name of Newton occurs three times in the *Crisis*, but only as a name. Galileo is Husserl's hero. It is a Galileo of the 1930s, perhaps best represented a little later in the work of the great French historian of science, Alexandre Koyré. That is, a Platonist Galileo who, to exaggerate, never did an experiment in his life but worked out the mathematical form of nature. No coincidence that Husserl and Koyré shared a common attitude to Galileo. Koyré, like so many other French humanists, was trained as a philosopher, and as a young man went to study with the best, namely in

Göttingen with Husserl. He was also an active member of Husserl's phenomenological circle in Munich.

Galileo the experimenter made a comeback in the 1970s. An indefatigable amateur, Stillman Drake, found Galileo's autograph records of experimental observations and confirmed that you would observe those results with Galileo's modest apparatus. Which is not to say that the vision of Koyré and his immediate predecessors was wrong, only that it was incomplete, although not as incomplete as that of Kant. Husserl's Galileo was above all the mastermind behind the mathematization of nature. What Husserl named "the Galilean style" of reasoning was the mode of reasoning by which, in Kant's words, "*physics* entered on the highway of science." The implied notion of a relatively small number of fundamentally distinct styles of reasoning has attracted a number of twentieth-century writers, from at least the time of Oswald Spengler. The cosmologist Steven Weinberg picked up the phrase Galilean style from Husserl, and Noam Chomsky picked it up from Weinberg.[2] For those readers, the Galilean style was the method of making abstract mathematical models of phenomena, and only later, when possible, testing them against experience. Husserl certainly meant that, but he also meant something more specific. He thought that Galileo not only made mathematical models of nature: He also transformed nature by making her geometrical.

No one now reads Husserl for a literal history of science. His is a rational reconstruction, and his Galileo is less the historical Galileo than an emblem of what happened to the sciences in the late Renaissance. The world became mathematical, or rather, in his opinion, geometrical. In Galileo's day, and for some time afterward, *geometrical* pretty well meant what we would call mathematical. But Husserl had a stronger reference, for he thought of Galileo as planting a geometrical structure on the world. And this is a matter of a vision of nature as composed of the ideal objects, the ideal structures, of geometry. This leads directly to the project of the *Crisis*. Physical scientists have lost touch with the way in which the ideal objects of geometry have come into being. Hence they have what is in effect a false image of what they are doing. I hope it is not inapt, or an abuse of overused phrases, to say that they work unwittingly in bad faith; theirs is the unhappy consciousness. Only a return to primal sources, to the Ursprung, will liberate them. In particular, we shall regain a full sense of the objectivity of the sciences by comprehending the source, the originary evidence, for this objectivity.

I have already expressed skepticism about this desire to uncover the origins as the deep source of objectivity. But there is a more specific worry. Galileo (the emblem that we call Galileo) did mathematize the world—by which I think Husserl meant that the new science caused the scientists of that and all later epochs to conceive of, and indeed to perceive, the world as having a structure described by and conforming to mathematics. But the world was not geometrized in the sense in which we now think of geometry. Galileo mathematized *motion*, a mathematics that culminated in the differential calculus of Leibniz and Newton. I am not at all sure that it makes sense to say that the calculus has ideal objects. Perhaps the Newtonian version does—the infinitesimals of which the good Berkeley made so much fun. Does the Leibnizian version have ideal objects? I think not.

This observation is not an instance of the carping that is to be avoided. Husserl was out to give a sense of the objectivity of mathematical reasoning in all its richness. In my opinion it would have served him well to attend to the new *proofs* that arose in the Galilean era and on into the epoch of rational mechanics. Readers more close to the spirit of Husserl will retort that I am thinking in too limited a way of the objects of geometry as circles and isosceles triangles. No, what was meant by "ideal objects" was ideal formal *structures*. Excellent. Let us compromise by saying that living proofs that can be grasped are the mathematics-in-action side of the Galilean style, and that the structures are what it produces.

Proof and Its Objects (Bis)

Kant saw physics as entering the royal road of science with the empirical methods of Bacon, Galileo, and Stahl. Husserl saw it as mathematizing nature. Husserl's vision of the new physics was a profound advance on that of Kant. Yet at another level their pictures of ancient mathematics are close kin. We have long been under the sway of the great German historians, and a few French and English ones, for whom the Greeks were indeed "that wonderful nation" that created geometry. Nowadays, as already stated, a critical rereading and rethinking of ancient texts is under way. There was more to ancient Babylon than the old historians taught, and there is a great tradition of Chinese mathematics. Alas, I have not yet been reconstructed. I think that the possibility of making proofs, the discovery of mathematical demonstration, really did take place on the shores of West Asia, by people who numbered among them Thales, whatever was his name. And I have no more problem in using "Thales," without the usual qualification, as the name of one such man. So I am closer to Kant and Husserl than more contemporary minds.

I am not so unreconstructed as to imagine that "a new light flashed upon the mind" of just one man, that man whom we call Thales. Maybe it did, but he had to create a school that grasped what he did. It is more likely that there was an emerging awareness of the possibility of proof, of gaining knowledge by thinking and arguing, of an independence of certain facts from the confirmations of measurement. Of course there could have been just one amazing figure who perceived the first proof, and who by talk and toil created a community in which mathematical proving went on. What we do know is that there was a long, slow development of proof ideas, and that during that process, proof was experienced as something extraordinary. Even very late in the day, Plato used the possibility of proof, illustrated by the slave boy in *Meno*, in an astonishing argument for the immortality of the soul. He insisted that decades spent studying mathematics was an absolute precondition for joining the administrative elite of a wise republic. Call Pythagoras and Plato math-obsessed freaks: Even so, no other civilization left traces of freaks like those who were also cultural icons. So I agree, in principle, with the vision of Kant and Husserl. Yes, there *was* a revolution as profound as any in the history of the human race, and it had a profound impact on all future Western philosophizing.

One hopes that postcolonial history of mathematics will teach us more about an emblem other than Thales, namely Al-Khwarizmi (c. 785–850). That name gave us the word *algorithm*. His book *Hisab al-jabr w'al-muqabala* gave us the word *algebra*. Among my as yet unreconstructed beliefs is that in the House of Wisdom in Baghdad, sometime around 850, say, they understood what an algorithm is. It is a procedure that by routine application of rules definitively solves any chosen member of a given class of problems. The *Nine Chapters* of Chinese mathematics mentioned above had a not unrelated concept six centuries earlier than the Arabians. But the Chinese commentaries immediately introduce "corrections"—that is to say, these are not strictly algorithms in the Arabic sense of the word, but systems of approximation. The Chinese rules are brilliant, but also exactly what an algorithm isn't.

A simplistic reader of Husserl might go on to urge that the Chinese rules therefore lack the objectivity of the mathematics descended from Greek and the Arab sources. Not so: They are systematic objective approximations. They are closer to the actual practice of physics as we know it than our official methodology that wrongly teaches that approximation is always a mere approximation towards the *real* truth. If world-historical mathematics had developed from the Chinese model rather than by a fusion of the Greek and Arab models, we might have had far healthier philosophies than the ones we have been saddled with. I do not just mean a richer philosophy of mathematics or of the sciences: I mean philosophy. But it would not have been a philosophy so curiously populated by those in awe of mathematics—Plato, Leibniz, Russell, and the rest whom I have mentioned. But note that, pace Husserl, our mathematics and our sciences would have been just as objective as the actual ones that that took the royal road of ideal objects.

The Historical A Priori

We should now consider Husserl's phrase "historical a priori." It is easy to be misled here, for this very phrase was to be lifted out of Husserl by a couple of generations of French philosophers. *Ursprung* was first published in the *Revue internationale de philosophie* in 1939. Ideas related to it and other aspects of Husserl's phenomenology of the sciences were analyzed and criticized by the brilliant young philosopher of mathematics Jean Cavaillès, since canonized for his execution as a resistance fighter. His work sparked the French revolt against phenomenology. It was in many ways at the heart of the French abandonment of their own Jean-Paul Sartre and Maurice Merleau-Ponty. Jacques Derrida translated *Ursprung*, but what he took from it was the sensible emphasis that one aspect of the transmission and indeed sedimentation of mathematics is the transcription of arguments in sentences. After all, we know about Euclid because it is written down. Derrida turned that elementary fact into the amazing doctrine that the sentence is primary and trumps the spoken word. The promotion of the pure sentence, stripped of speech and of both speaker and author, accompanied—may even have led to—the "death of the subject," rumors of which have been greatly exaggerated. Nevertheless, some of Husserl's phrases stuck.

Notable among them was historical a priori, used, with a mixture of high seriousness and mordant irony, by Georges Canguilhem and Michel Foucault. The passage from Husserl through Cavaillès and Canguilhem to Foucault has been well chronicled and analyzed by David Hyder.[3] I myself have traveled on the edges of this awesome field of force. Hence you may find in my discussion some of what Howard Bloom called the anxiety of influence.

What did he mean by historical a priori? I have heard it said that he meant no more than truths about history that we can know a priori—for example, that geometry must have originated in such and such a way. Given the present tenor of history of mathematics, that is a rather uncharitable understanding of Husserl. We have also heard more complex explanations, which tie it in to the entire project of transcendental phenomenology. I would like to propose a simpler account.

The phrase occurs only in the *Ursprung*. So it is plausible to suppose that it has a specific connection with geometry and its origins. I suggest first that Husserl is speaking of a priori knowledge in exactly the sense of Kant. *Except* that Kant thought that structures knowable a priori were preconditions for possible experience, at all times and places. Those structures are out there, to speak crudely, and the man whom we call Thales just happened to pull the light switch that was always there on the wall, always available to illumine some mind. In Husserl, the light that flashed upon the minds of the first mathematicians was a historical event before which there was *no* mathematical, geometrical, a priori. Thus I propose that historical a priori denotes *an a priori structure that comes into being at a historical time.* Perhaps the chief merit of this reading is simplicity. It is not so relativistic a thought as might appear: We should recall that a priori is an adjective whose first job is to characterize a type of knowledge. Even if there are eternal truths, *knowledge* of them comes into being in history.

To classify geometrical knowledge as historical a priori is not to imply that all a priori knowledge is historical in the same way. One imagines that a priori knowledge about geometrical objects was preceded by other types of a priori knowledge dependent upon analytical connections; perhaps also there is a priori knowledge dependent upon aspects of experience that phenomenology can reveal.

My understanding is a "French" one. When Canguilhem and Foucault spoke of the historical a priori, with a touch of irony I am sure, they meant the structures of systems of thought that determine the space of possibilities of what can be said. Not exactly preconditions, but conditions of possibility and exclusion that come into being, and may go out of existence, at canonical moments. In Foucault's case, those moments tend to coincide extensionally with 1619, namely the year of Descartes' night in the *poêle*, or 1789, *the* revolution. Yes, such historical a prioris can mutate or cease to exist, but they are otherwise Husserlian in that they are (a) historical, coming into being in historical or near-historical time, and (b) a priori determinants of all possible surface knowledge in a domain.

Sedimentation

Husserl gives a very vivid account of the ways in which the original insights and achievements of the first mathematical communities might (or must?) have been developed into written texts. They evolved into Euclid's *Elements* and grew and grew, into Galileo and beyond, an ever-opening horizon of new mathematics. As layer upon layer of proofs and texts evolve, so we have more and more sedimentation. And it becomes increasingly time-consuming to dig down right to the original evidence of what we call Thales's generation. Descartes had something like the same thought. He held that you should some time in your life become convinced of every foundation (recall that Descartes was "foundational" in a way that Husserl never was), but you cannot do that every day of your life. Be content to remember that you once had a clear and distinct proof before you. The era of Descartes—that of Galileo—was just the beginning of the Husserlian mathematization of nature. Endless layers of proofs and postulates have piled up since. The metaphor of sedimentation is the most natural one in the world. And whereas at the time of Descartes and Galileo, one really could reach down to bottom at least once in one's life, that is no longer practicable for the working mathematician.

There is a related, and perhaps more primal, kind of sedimentation. Husserl insisted that long before the man whom we call Thales, there must have been experience of approximations to ideal objects. Boards had to be made straight, planed, in order to be useful lumber for building. Lumber? Better perhaps to say that blocks of stone were hewn, and bricks baked as rectangular solids, so that they would fit together and pile up. Thus the empirical straightedge came into being. We should be morally certain that the intuition of the first circles came from the potter's wheel.

There was also measuring and surveying. There were lines of sight. You might think that the line of sight is an abstraction, a taken-for-granted that would never be experienced, intuited, or named. That is armchair speculation; anyone who has ever surveyed in hot and sandy country knows of a terrible problem: Heat makes things shimmer, so you cannot get a decent sight line. And so you get the idea of an ideal line of sight during the cool of the clearing dawn, and finally idealize that into the straight line.

Husserl seems to have believed that the sedimentation of both concepts and proofs, so necessary for any practical purposes, hindered anyone in the modern world who wanted to delve for the Ursprung of geometry, and hence to grasp the origin of all science. That is both origin in the sense of beginning and origin in the sense of primal evidence. For those who take this to be a serious problem—for those who lust after the primal as I do not—I propose a reality check, first into pure mathematics and then into mathematical physics.

Sedimentation in Pure Mathematics

Real, creative, pure mathematics is far less sedimented than Husserl seems to have thought. It is true indeed that textbooks looks like middens. But go to a research talk by a mathematician. No notes. No sentences tediously displayed by PowerPoint in an authorized Microsoft format. Quite often some drawings, perhaps with a felt pen on a transparency to save time—only the material differs from the drawings that Socrates made in the sand for the benefit of Meno's boy slave. Since Husserl speaks of geometry, I shall give two geometrical examples, one personal, one public.

I wanted to understand the construction by ruler and compass of the seventeen-sided regular polygon. Gauss discovered how to do it when he was eighteen or so. He also showed that the regular heptagon cannot be constructed with ruler and compass. A truly Euclidean problem solved. I obtained a rough idea by consulting sedimented sentences, including virtual ones on the Web. But I did not fully grasp the proof idea. I asked a mathematical colleague for help—it was just before the beginning of a rather dull, but very long business meeting. I saw him across the table, beavering away for almost two hours. At the lunch break, he had a couple of pages of scribbles every which way, to which he occasionally referred when explaining the idea. He had literally gone back to "first principles." Not by going back through a classic text (Hardy and Wright's 1938 *Theory of Numbers* would still be a natural, if sedimented, choice). There were certain things he took for granted—the idea of the circle and the fact that it is defined by a quadratic equation. So anything you do with a ruler and compass can be described by quadratics . . . now start thinking of all the possibilities . . . a certain group of transformations looms up before you. . . .

My friend said with a certain feeling of awe, "You see, Gauss hit upon the rudiments of the theory of groups before he was eighteen." I suggest this personal story as a parable for pure mathematics in action. If as a philosophical outsider you examine textbooks, the notion of heavy sedimentation is immediately forced upon you. But if, even as such an outsider, you become a participant observer in mathematical activity, you will find very little sedimentation. There is the living experience of the very evidence, the primal evidence, that Husserl sought.

My second example comes from the popular science press, accessible to anyone.[4] It is also geometrical. It is the story of the late David Huffman (1925–99), a distinguished computer scientist who became interested in a more sophisticated version of the construct-with-ruler-and-compass problem. What shapes are constructable by folding paper? That is, what is the class of all possible origami? People who know about the calculus of variations will know there is indeed a history here. One stage goes back to the blind Belgian mathematician and physicist Joseph-Antoine Plateau (1801–83). He determined the surfaces of least area subtended by a given closed curve in space by telling sighted assistants to dip wire shapes in a soap solution and tell him the resulting soap surface. Then he went on to demonstrate the solutions for special cases.[5] Aspects of Plateau's problems continue to excite interest, and there are still things to find out. That is where Huffman started his career, but he moved on to folding paper, also allowing himself curved paper not used in classic origami. And so he went on to develop remarkable abstract theorems, *from scratch*, as we say. (For example, the pi theorem: If you have a point surrounded by creases, and you want the form to fold into a plane, all alternate angles must sum to pi radians, or 180 degrees.) Popular science accounts, such as the one cited, include photographs of quite amazing shapes that David Huffman showed how to construct from a single sheet of paper. Even what Husserl would call the ideal shapes, the ideal objects that Huffman created, were new. Eternal possibilities, you may say, but historical a priori realities that came into being fairly late in the last century and exist on a horizon of possibilities that are being investigated at this very moment. Amount of sedimentation? Practically zero. Another parable.

Sedimentation in Mathematical Physics

Throughout the week of June 28 through July 3, 2004, all the string theorists in the world met in the amphitheater two floors below my office in Paris. I know even less string theory than I know Husserl, but I took the opportunity to sit in on a few talks early in the week. It was a fascinating experience in the anthropology of science, but here I restrict myself to one observation. It was astonishing to see that a great deal of the mathematics being used was classical mathematical physics of the sort that firmed up in the nineteenth century: Lagrangians, Fourier transforms, tensors, slightly more recent matrix theory. It was pretty much what I learned in my third or fourth year as an undergraduate at a mediocre department of mathematical physics nearly fifty years ago. The applications were radically new speculations and structures, for sure. I am referring to the tool kit employed by these (mostly very young) string theorists.[6] It is very natural to use the metaphor of sedimentation to apply to what, using another metaphor, is the tool kit of cutting-edge mathematical physicists.

In the *Crisis* Husserl was indeed addressing the most abstract sciences of his day, above all mathematical physics. If you did not pause in your philosophizing to *take a look*, you would only suppose that there has been more sedimentation since 1935. But here we have a natural science whose sedimentation appears to the amateur ethnographer to be not much different from that of Husserl's time. String theory a natural science? Well, a speculative science that aims at a deeper understanding of the physical world than ever was before.

All right, it will be replied, but still Husserl was correct. Even if the sediment used in the tool kit of the string theorist was mostly known to Max Planck, it was still sediment! Yes—but sediment not only easily excavated, but excavated before your very eyes when you are a youth learning the trade. One of the reasons that the tools used in these research reports were immediately recognizable to me was that once upon a time, I was shown where they come from, I was taught the ideal version of the fundamental evidence on which they were based. No, I was not literally given a course in then history of mathematics, which would tell me just how Stokes and Green hit on their theorems. Not an Ur-experience like that, thank goodness. But when I got home, I checked in my still-saved textbook of long ago (sedimentation once again) that I had been shown, more or less from scratch, the evidence for Stokes's and Green's theorems and could (in a manner more reminiscent of Descartes than Husserl) pretty well relive that experience right now. And that despite the loss of the relevant bits of brain capacity over the years. Happy coincidence: The text was written by Richard Courant, who had been Hilbert's assistant in Göttingen in 1907. Right alongside Husserl who worked there 1901–16 (Courant, 1937). And Courant also wrote that classic exposition of Plateau on soap bubbles.

In short, I suggest in this section and in the previous one that there is far less sedimentation than armchair inspection of mathematics and physics textbooks would suggest. Oddly, I think that Jacques Derrida was faithful to an aspect of Husserl when he took away the thought that the sentence is primary, trumping speech. The trouble is that Husserl himself, when he had become elderly, seems no longer to have *taken a look* at reason in action. He thought we were separated from the original evidence by almost undiggable layers of sediment. I said in the Primal Beginnings section above that the primal does not seem to be so important as Husserl thought. But even if it were, I have argued in these last two sections that we are far closer to innumerable primal *Ursprünge* of geometrical evidence than Husserl noticed.

Cognition, History, and the A Priori

I will seem to have distanced myself so far from Husserl, that, contrary to the praise expressed in the Husserl aux pantoufles section above, I have come to bury Husserl on geometry. Not so. I resonated with his (what shall we call it?) attempt to weld together historical speculation and speculative philosophy. My own such ongoing attempt, of which only a threadbare sketch was published over a decade ago,[7] holds that mathematical reasoning and demonstrative proof are a discovery of human civilization at the dawn of history. The emblematic, mythical origins of what we now call mathematics include the man whom we call Thales, truly lost in the mists of the backward-looking horizon named prehistory, and the man whom we call Al-Khwarizmi, whose historical past will probably be reconstituted with far more detail in the next few years as scholars turn their eyes from the Greek to the Islamic mathematical tradition. Both the geometrical and the algorithmic (algebraic, combinatorial, what you will) styles of reasoning—which united but did not merge in the European Renaissance—originate in the history of particular human societies, perhaps of the Ionian coast and of Baghdad, but they rely on universal human capacities. That is, they are a phenomenon that emerged in culture, but which are possible only because of cognitive universals. In Kant's myth, Husserl's myth, my myth, Thales experienced the human potential to produce that engine of discovery, mathematical proof and the associated ("ideal") structures.

I hold the subsidiary thesis that every style of reasoning (in the specific sense that I have developed elsewhere [1992]) introduces its own domain of new objects. That is why there are ontological debates—over the reality of mathematical objects, over the reality of unobservable theoretical entities, over the reality of the taxa of systematic biology, and so forth. So my view of the relation between ideal objects and their production by demonstration is Kantian, not Husserlian. But even though I hold that ontological thesis to be of fundamental importance in diagnosing the source of realism/antirealism debates of the sort that continue to flourish among philosophers, it is of no great importance for this chapter. What does matter is that the legend of Thales meets a different precondition than that upon which Husserl focused. The precondition is not a historical a priori but rather a cognitive one. It is part of our human genetic envelope (a phrase I take from my colleague Jean-Pierre Changeux). It is the kind of thing about which some cognitive scientists like to speculate, when they imagine that we have one or more innate cognitive modules that enable us to reason mathematically. Much such theorizing consists in hand waving—we know a good deal less about cognitive modules for reasoning than we do about the day when the "new light . . . flashed on the mind of the first man (*Thales*, or whatever may have been his name)." But I want to encourage more hand waving, not less. Could it, for instance, be that diagrams were not mere adventitious tools that helped in explaining geometrical arguments? Could it be that only through the use of diagrams were humans able to "access" those regions of the brain that are essential for constructing proofs (Netz, 1999)? There opens up a horizon of research, part historical and cultural, part neurological and cognitive. It continues the most important of the reflections that Husserl began in his lecture of 1935, on the *Origin of Geometry*.

§ 5 The *Crisis* as Philosophy of History

David Carr

The *Crisis* texts have had a checkered history and a widely varied reception. As many readers will know, only parts of the original book were published in an obscure exile journal in Belgrade during Husserl's last years. The real history of its influence began when its unpublished parts, smuggled out of Germany by H. L. van Breda, were consulted by the young Maurice Merleau-Ponty at the newly founded Husserl Archive in Leuven in 1939. The long Part III of the *Crisis*, together with the unpublished *Ideas II*, were the texts that impressed Merleau-Ponty the most. In Husserl's concept of the life-world, he found support for the existential version of phenomenology, strongly influenced by both Sartre and Heidegger, that he presented in his thesis, *Phénoménologie de la perception*, published in 1945. It wasn't until 1954, with the posthumous publication of the more or less full text of the *Crisis* in the Husserliana edition, that the larger philosophical audience was able to get beyond the somewhat slanted interpretation imposed by Merleau-Ponty to see the concept of the life-world in its original context, and to get a proper sense for its role in the development of Husserl's thought. Nevertheless, the *Crisis* was still read primarily as a document in the history of continental philosophy, as that term was used throughout most of the last half of the twentieth century. As someone with a long association with this text, I am pleased to see that the *Crisis* is finally being taken seriously as a contribution to the philosophy of science, which is clearly one of the things Husserl intended it to be.

But that is obviously not all he intended. Husserl thought of the projected work as a final, comprehensive statement of his philosophy as a whole, a sort of philosophical testament, and there are many facets to it. Since I am not a philosopher of science, I will leave it to others to assess the work from that point of view. Instead I shall consider it primarily in my capacity as a philosopher of history. Emphasizing the life-world, Merleau-Ponty tended to portray the *Crisis* as Husserl's deathbed conversion from transcendental philosophy, or "intellectualism" as Merleau-Ponty called it, to existentialism. But it seemed to me from the start of my own acquaintance with this text that its preoccupation with history was what was genuinely novel about it, at least in comparison to Husserl's earlier work. I addressed the theme of history in Husserl's late work some years ago; now I want to return to it, after some long detours, and examine this theme again.

What I want to ask is: How is this preoccupation with history to be understood? Does it constitute a philosophy of history, and if so in what sense? How does it compare with other approaches to history? Husserl's approach is, I think, very hard to classify. There are many aspects to this preoccupation with history, as many readers will know. He advocates a historical approach to epistemology in general and to the philosophy of science in particular.[1] He deals with the history of science and of geometry; he devotes a long discussion to the history of philosophy; and he tries to develop the "philosophical-historical idea of Europe" (*The Vienna Lecture*, p. 269). He puts forward some very interesting but extremely difficult reflections on his own historical method of investigation. Does any of this qualify as philosophy of history in any recognizable sense?

It is customary to distinguish between the substantive or speculative philosophy of history on the one hand and the critical or analytical philosophy of history on the other. The former is supposed to belong to metaphysics, making claims about the historical process itself, and the latter to epistemology, since it is about historical knowledge. This distinction, which dates to the analytic philosophy of the 1940s and 1950s, was of course unknown to Husserl; it was made in order to establish the analytical philosophy of history as a respectable enterprise and to put the speculative form, along with other forms of speculation, and indeed of metaphysics, out of business. For this and other reasons, I think this distinction may have outlived its usefulness, but it may be helpful to examine Husserl's text in light of it. There are in fact elements of both approaches to history in the *Crisis*; but its most interesting contribution to the philosophy of history, I believe, belongs to neither of them. It is found, rather, as I shall try to show, in the concept of *Geschichtlichkeit*, or historicity.

Substantive Philosophy of History

Let us first consider the *Crisis* texts as substantive philosophy of history. Is Husserl advancing claims about history itself, in the manner of the classical philosophies of history? One's first impression, I think, is that this is exactly what he is doing. He seems to be attributing a teleological structure, a direction, to history as a whole, in the manner of the grandiose nineteenth-century theories of Hegel, Marx, or Comptean positivism. From his Viennese background, Husserl had inherited a certain disdain for Hegel, mentioning him rarely, and he seemed genuinely uninterested in the fact that Hegel made important use of the term *phenomenology*. As far as I know, he never mentioned Marx. The positivist link is more plausible, since we know of Husserl's connection with the German branch of Mach and Avenarius. And yet the figure who looms largest here is Hegel, who was in any case the paradigmatic philosopher of history in the modern period.

It is easy to find echoes of Hegel's philosophy of history in Husserl's text. It is in Europe that humanity has really come into its own. Europe, Husserl assures us, is not a geographical expression but an idea, and while the citizens of the United States and of the "British Dominions" are accorded the status of honorary Europeans, the Eskimos, Indians, and Gypsies who inhabit those precincts are not. Other non-Westerners, such as the Chinese and "Negroes in the Congo" are mentioned, not necessarily disrespectfully, but clearly as outsiders. The term *Geist* occurs frequently in this work. Husserl's ideal Europe begins with Greek philosophy and culminates in the present. He employs the term *teleology* copiously to describe this trajectory, from the *Urstiftung* (primal establishment) of European humanity in Greece to the *Endstiftung* (final establishment) embodied in phenomenology itself. This culmination is viewed by Husserl as the triumph of reason in history.

There are obvious dissimilarities, of course. Husserl's description of this trajectory is, if anything, even more idealistic than Hegel's. For Husserl, history is driven by or even consists entirely in the progression of ideas. There is no attempt, as there is in Hegel, to relate political and cultural forces to these ideas. The drive toward freedom, which is central to Hegel's philosophy of history, does not figure prominently in Husserl's account. The dialectical movement, the cunning of reason, the role of human passions—all those flourishes that are distinctive of Hegel's account of the course of history seem to be missing here. Reason alone, in the form of *Wissenschaft*, or science in the wide sense, is the solution to humanity's problems; and we find no positive role for religion here either. In this respect, Husserl seems closer to the Enlightenment conception, where reason overcomes the forces of superstition, or to its later, positivist counterpart, where science (in this case phenomenology) supplants both religion and metaphysics.

But the larger contours of Hegel's system seem to find their counterparts in the *Crisis*. The Greeks discover the distinction between reality and appearance, nature and convention. Man moves toward the outside world, captures it in modern objectivism, meanwhile forgetting the subjective accomplishment that made it possible. The stage is set for a return to subjectivity, in the transcendental phenomenological turn. Spirit others itself, returns to itself, in the familiar Hegelian pattern. History is the story of that departure and return. Thus we are being told that history has a meaning, purpose, direction, in the manner of the classical, speculative philosophy of history.

Though Husserl gestures in the direction of such a philosophy of history, it would be a great mistake to read him in this way. In fact, this classical account of history, which places its hopes for the salvation of mankind in reason, philosophy, and science, is evoked by Husserl, in the early pages of the *Crisis* and elsewhere, as an object of bittersweet nostalgia and with a sense of loss. The European sciences trace their origin to a time when these ideas could still be taken seriously, when knowledge was supposed to make us wise and give meaning to life. But now they have been separated from each other, from the guiding ideal of unity represented by philosophy, and above all from the ordinary human life to which they were supposed to give meaning. This is the crisis of European sciences: the loss of their meaning for life.

The term *crisis* is the telltale sign that Husserl is not presenting a Hegelian-style philosophy of history. The idea of crisis has no place in the Hegelian scheme. Hegel assures us that reason has triumphed, or is about to. We have emerged from the excesses of the French revolution and the Napoleonic wars and can construct constitutional states that achieve the freedom promised by the Enlightenment while preserving the best of the ancient monarchies. Even religion, which the Enlightenment wanted to sweep aside, can find its place in the new order. The crisis is over. For Marx, the revolution is not a crisis; we know how it will turn out.

Again Husserl may seem to be closer to the milder form of the substantive philosophy of history associated with the Enlightenment and with Kant. The French *philosophes* believed in progress and affirmed that it could be achieved through human agency; Kant very much wanted to believe in it, but with his usual caution argued only that it could not be ruled out and thus could legitimately be hoped for and, especially, striven for.

But here too the idea of crisis has no place. We can distinguish three different narrative strategies, as we might call them, in the modern substantive philosophies of history. Hegel and Marx give us closure, a fairly clear-cut End of History to go with its beginning and its middle. The Enlightenment's future is still open, but the idea of human salvation is pretty clearly implied, even if we cannot give it a full-fledged definition. Kant thought a league of nations might do it. But the idea of crisis places us in the middle of a fateful drama, at a turning point where the possibility of a reversal of fortune looms large before us. The metaphor is medical, of course: The patient is ill; things could go either way. Something must be done. Human agency is called for in all three of these models, even the Hegelian-Marxist one, though it is often portrayed as deterministic. But in the case of a crisis, the need is urgent: Emergency intervention is called for.

Husserl's choice of metaphor may seem entirely apt, given the situation he was in, lecturing in Vienna and Prague in the years leading up to the *Anschluß*, or annexation, and the Munich conference. But in fact, as Charles Bambach points out, the term "*Krisis*" figured in the titles of several much earlier studies Husserl probably knew about: Rudolf Pannwitz, *Die Krisis der Europäischen Kultur* (1917); Ernst Troeltsch, "Die Krisis des Historismus" (1922b); and *Die Geistige Krisis der Gegenwart* (1923) by Arthur Liebert, the same man who later edited the journal *Philosophia* in Belgrade, where Husserl was to publish his own *Krisis* text. And Heidegger had spoken of the crisis of the sciences in his lectures of 1925, and again in his 1927 *Sein und Zeit.* Husserl indeed admits, at the beginning of the Vienna lecture, that the theme of the European crisis has been much discussed. Clearly the "crisis" as a historical topos belonged to the whole interwar period, at least in central Europe, and tells us a lot about how its intellectuals thought about what they were going through. This third narrative strategy, typical of the early twentieth century, and even beyond into the cold war period, was lacking in the nineteenth.

Still, this book is Husserl's crisis in more senses than one. Clearly he thinks the fate of German philosophy, as he conceives it, hangs in the balance. His defenders after World War II were eager to deny Merleau-Ponty's interpretation by showing that he never let go of his ideal for philosophy and that he did not himself assert that the "dream" of "Philosophy as Rigorous Science" was exhausted, as some passages seemed to suggest. Yet the poignancy of the text as a cri de coeur, its tone sometimes close to despair, is unmistakable. And indirectly, as we know, it is deeply personal. It is not only that Husserl identified himself completely with philosophy. Like many of the assimilated Jews of his generation, he considered himself a loyal champion of Western culture and a citizen of Europe. He had converted to Protestantism at an early age, under the influence of his youthful friend Gustav Albrecht and his mentor Thomas Masyryk. By all accounts, he took it seriously: Husserl took everything seriously, and it is almost impossible to think of cynical or self-interested motives in his case. He seems to have focused on the ethical teachings of Jesus in the New Testament. Later he was a conservative, probably a monarchist, and he bore the archaic title *Geheimrat* (privy councillor) with great pride. One of his sons died on the field of battle in the First World War. In a bitter disappointment, his most gifted follower, who once addressed him as "*lieber väterlicher Freund*," (dear fatherly friend) had deserted phenomenology and was now representing the forces of irrationalism as a philosophical storm trooper. His dismay at what was happening was echoed in the later testimony of other Jewish intellectuals, of his sons' generation, like Karl Löwith, who lost one of his lungs for the Fatherland, and the famous diarist Viktor Klemperer, a professor from Dresden and another veteran of the first war.

These personal themes, inextricable from the text of this work, give a special flavor to Husserl's historical reflections. Nevertheless, insofar as they are philosophical reflections on the course of European and even world history, they can be considered as belonging to the substantive philosophy of history. But as I have already suggested, they are far indeed from the themes and metaphysical claims we associate with the classical models of the eighteenth and nineteenth centuries.

Critical Philosophy of History

So far we have examined Husserl's text against the background of the substantive philosophy of history. Can it also be regarded as a contribution to the critical philosophy of history? Husserl can indeed claim some credentials as an epistemologist of historical knowledge, at least indirectly. Though the analytic philosophers of the 1950s tended to ignore it, the epistemology of history didn't begin with Carl G. Hempel. Debates about the status of historical knowledge began at about the same time, in the nineteenth century, that history established itself in the academy and historians began to claim that it was a Wissenschaft, rather than merely an entertaining and edifying literary genre. Members of the "historical school" (Ranke, Niebuhr, Droysen) had a lot to say about critical methods for evaluating sources, interpreting documents, and justifying their assertions. But the larger question for philosophers, of course, was how the newly flourishing historical knowledge related to that of the natural sciences, which had served as the paradigm for epistemology in the seventeenth and eighteenth centuries. The positivist tradition, inaugurated by Comte and Mill, argued for the unity of all knowledge and tried to assimilate history to science. Just as physics formulated laws of nature, and explained events by their means, so the science of society would seek out social laws; history was just a case of applying these laws to the past.

Led by the neo-Kantians (Windelband, Rickert) and by Dilthey, German philosophers opposed this view of historical knowledge, focusing on the fact that its objects are not natural occurrences but human actions. It was with history in mind that they began to work out the idea of the *Geisteswissenschaften* or humanities, maintaining the idea of the autonomy and independence of disciplines concerned with human affairs against the attempt to reduce them to something more basic. This opposition between the positivists and the humanists continued to shape the debates about the status of historical knowledge well into the twentieth century.

Husserl always sided with Dilthey and the neo-Kantians on the matter of reductionism; but he wanted to work out the distinction between the *Natur-* and the Geisteswissenschaften, the natural and the human sciences, on his own phenomenological terms. Given his differences with the neo-Kantians, he drew more heavily on Dilthey, whom he regarded as a man of "ingenious intuition," but unfortunately not of "rigorous scientific theorizing" (*Ideas II*, p. 173). Husserl began his work on this topic in the studies for the second volume of *Ideas*, by developing the idea of constitution. Though he did not deal with historical knowledge directly, he did concern himself with the difference between knowing objects in nature and knowing persons and understanding and interpreting their actions. He developed the idea of Natur and Geist as distinct ontological regions, each with its own material a priori repertoire of concepts, determining basic entities, principles of individuation, and relations of temporality, spatiality, and causality. Moreover, he conceived of the area of transition between these two traditional realms as a distinct region of its own, that of "*animalische Natur*," or *Seele*, where animals and humans shared certain bodily properties, sensations, capacity for movement, and rudimentary intentionality. His investigations here on the lived body, or *Leib*, as "center of orientation," as bearer of will, movement, and habit, and of visual and tactile intentionality, served as Merleau-Ponty's inspiration, and these and other unpublished manuscripts of the period are coming to be recognized as surpassing in subtlety and sophistication those of the French philosopher who later appropriated them.

To each of these regions belongs, on the side of the observer-scientist, a distinct *Einstellung*, an attitude or frame of mind that brings to the experience of each domain certain basic concepts, expectations, and forms of inference. The general "natural attitude" of *Ideas I* is now subdivided into the "naturalistic attitude" underlying the natural sciences, the "personalistic attitude," corresponding to the Geisteswissenschaften, and a third, which Husserl does not name, that underlies the science of psychology. One of the most important discussions here concerns the distinction between causation in the natural realm and motivation in the human world. The concepts of *Umwelt* and *Welt*, environment and world, also play an important role in the discussion of persons.

Considered as contributions to the epistemology of the Geisteswissenschaften, these discussions, from the 1920s and shortly before, are certainly relevant to the philosophy of history. However, not much of this turns up in the *Crisis* itself. In fact, some of the subtlety and detail of Husserl's manuscripts on these subjects, especially on the distinction between Seele and Geist, gets lost during the late period, when Husserl develops the idea of a phenomenological or intentional psychology mainly in order to discuss its relation to transcendental phenomenology. This is the subject matter of Part III B of the *Crisis* text, which takes its point of departure by criticizing Kant and his successors for misunderstanding the distinction between psychology and epistemology. Husserl wants to advance the idea that if intentionality is pursued to the limit, a psychological investigation of consciousness, properly understood, can transform itself into a transcendental philosophy with a mere "change of sign" of the sort brought about by application of the full-fledged phenomenological reduction. An intentional psychology can thus function as a way into transcendental phenomenology—a problem that concerns Husserl during this period. He explicitly criticizes the Cartesian approach of his earlier works and is looking for alternative ways of presenting his method. These sections of the *Crisis* are thus devoted to two related problems that come up elsewhere in the work: first, what he calls the "paradox of human subjectivity: being a subject for the world and at the same time being an object in the world," (*Crisis*, § 53) that is, the problem of transcendental versus empirical subjectivity; and second, the status of phenomenology as a "first philosophy" or self-sufficient philosophical method.

As so often happens in Husserl's programmatic texts, the detail and subtlety one finds in the manuscripts are sacrificed to large-scale methodological issues. What is missing here is a treatment specific to the Geisteswissenschaften, of the sort that might include history, as well as the specifically epistemological interest that guided Husserl's studies on constitution. As a contribution to the epistemology of the human sciences, the value of the *Crisis* is limited.

Historicity

As a contribution to the philosophical reflection on history, however, the chief value of the *Crisis* lies in another direction. Underlying the whole approach of these texts is a concept that Husserl employs frequently during this period, namely that of Geschichtlichkeit, or historicity. As we know, this term also figures prominently in Heidegger's work as well. An important late chapter in *Sein und Zeit* bears the title "Zeitlichkeit und Geschichtlichkeit." In fact, since *Being and Time* was published in 1927, and Husserl's work on the *Crisis* dates from the 1930s, it is possible that Husserl picked up the term from Heidegger, in spite of his negative feelings and very critical attitude toward Heidegger's work. A likelier story is that the importance of this term is testimony to the influence of Dilthey on both Husserl and Heidegger. Though Dilthey had died in 1911, the seventh volume of his collected works, which contained the author's late manuscripts on the *Aufbau der geschichtlichen Welt* (the construction of the historical world), was published in 1927. Heidegger explicitly pays homage to Dilthey at the beginning of his chapter on historicity. Husserl was acquainted with Dilthey's late work through his assistant Ludwig Landgrebe, and through Georg Misch, both of whom published studies on Dilthey from a phenomenological perspective. Though Husserl had been critical, in "Philosophy as Rigorous Science," of the historical relativism he saw in Dilthey's work, he later paid him tribute as a theorist of the Geisteswissenschaften and for his attempts to found a humanistic psychology.

A clue to understanding the concept of historicity, and its relation to the epistemology of the human sciences, is found in a passage from Dilthey's *Aufbau der geschichtlichen Welt.* "We are historical beings before we become observers of history," Dilthey writes, "and only because we are the former do we become the latter" (1968, 277–78). For both Husserl and Heidegger, the concept of historicity is the result of elaborating on what it means to say that we are "historical beings." The epistemology of historical inquiry gives way to an account of the historical character of experience and existence—*Bewußtsein* for Husserl, *Dasein* for Heidegger.

But Husserl's actual development of the concept of historicity, whatever it may owe to Dilthey or even to Heidegger, is rooted ultimately in his own earlier work. Specifically, it derives from his treatment of temporality and of intersubjectivity. As we know from the lectures on internal time-consciousness, consciousness at any level, whether perceptual, imaginative, or conceptual, whether passive or active, is a temporal flow with a retentional-protentional form. The present is experienced against the background of a past and in anticipation of a future. Whatever its intentional objects may be at any given moment, the intentionality of consciousness takes in the past and future of those objects, of the world, and of itself. The subject is not a substance persisting through time, or a timeless ego hovering outside of time, but a self-constituting synthesis of temporal relations. As Husserl later developed his notion of genetic phenomenology, he portrayed consciousness as a process of accumulating abiding convictions and habitualities, building up a sense of world and of self. "The ego constitutes itself for itself, so to speak, in the unity of a *Geschichte*," as Husserl writes in the *Cartesian Meditations* (§ 36). Here

Geschichte can be taken in the sense of an individual story or narrative of one's own life, rather than history in the usual sense. Similarly, Dilthey had compared the constitution of self to the composition and constant revision of an autobiography.

History proper enters the picture with the intersubjective dimension of consciousness. Though Husserl is often faulted for his treatment of intersubjectivity, notably in the Fifth Cartesian Meditation, there is no doubt that he considered subjectivity and intersubjectivity to be essentially interrelated. Like intentionality and temporality, intersubjectivity is an essential dimension of experience. It is not as if the subject could somehow exist alone and then encounter others. Intentionality is a perspective or point of view upon the world, and intersubjectivity is the encounter and interaction with a point or points of view that are not my own. Husserl's brief appropriation of Leibniz's concept of the monad, in the Fifth Meditation, though it is ultimately misleading and inappropriate, I think, is meant to portray the subject as an element in a vast interplay of points of view in which the objective world is constituted.

Husserl's account gets much more interesting when he goes beyond the abstract Leibnizian scheme, still in the Fifth Meditation, and conceives of intersubjectivity in the form of concrete communities. He speaks of *Vergemeinschaftung der Monaden* and coins the expression "personalities of a higher order" to describe such communities of monads (*Cartesian Meditations,* § 55). In later manuscripts this is further spelled out as "we-"intentionality, where the first-person point of view, inseparable from phenomenology, is shown not to be limited to the first-person singular; it can be exemplified in the first-person plural as well. The background of the past now becomes that of the social or intersubjective past, which now belongs to the individual subject by virtue of membership in a community.

Thus our expanded view of consciousness now includes history, so to speak, as part of its makeup. That is, the social past figures as background of individual and collective experience. And it does this prior to and independently of any cognitive interest we might take in the past or even any instruction we might receive about it. This is what it means to say, in Dilthey's words, that we are "historical beings": We are historical beings because we are conscious beings. While the basic elements of this conception of historicity were already in place, it is left to the *Crisis* texts themselves to develop them and to draw out their implications. As we shall see, some of these implications have problematic consequences for Husserl's idea of phenomenology.

But we should pause at this point to consider the status of this concept of historicity. Since it concerns not historical knowledge but historical being, it is clearly not epistemological but ontological. This does not place it back in the realm of the traditional, substantive philosophy of history, however, since it is not about the being of the historical process as a whole, but about the being of the subject. But to call it ontological is not quite correct either, at least in Husserl's usage, since the point here is not to describe persons in their material ontological region of reality, as they might be treated in the Geisteswissenschaften, but to describe consciousness phenomenologically. To put it another way, to say we are historical beings is not merely to say we are *in* history, that we arrive on the scene and then disappear at a certain point in objective historical time. It is certainly true, not only that each of us is such a being, an empirical ego, but also that we are aware of ourselves as such. But historicity is a feature *of* our *awareness itself*, our awareness not only of ourselves but of everything else as well. Indeed, historicity is a feature of *transcendental* consciousness, as Husserl uses that word. Whatever the term may have meant for Kant, for Husserl transcendental means world-constituting, world-making, world-engendering—though not, of course, world-creating: Only God can do that! Hence the "paradox of human subjectivity," mentioned earlier, of being both a subject for the world, transcendental subjectivity, and an object in the world.

There is no doubt that the growing importance of historicity, in Husserl's late work, is evidence of the increasing concreteness with which he conceives of consciousness. But Husserl's conception was always more concrete than was generally recognized. We have already mentioned his studies of embodiment, which date back to *Ideas II*, and which also reappear in the *Crisis*. Merleau-Ponty, and Sartre as well, were quite right to see and appreciate this aspect of Husserl's work. Husserl never had a problem with intentionality's being instantiated in a particular medium or a particular situation. It is for this reason that it can be bodily as well as *mental* in the usual sense of that word. This is also why its subject can also be the plural *we* as well as the singular *I*. The fact is that for Husserl the world is constituted by an embodied and historically situated transcendental subjectivity. Historicity, then, is not an ontological concept, at least in Husserl's sense, but belongs strictly to phenomenology, indeed transcendental phenomenology.

We have seen that historicity is not an epistemological concept, but it does have epistemological implications. When Dilthey remarked that we are historical beings, and because we are historical beings we become observers of history, he was saying something about historical knowledge. But he was not addressing standard epistemological questions about grounding, validity, objectivity, evidence, and so forth. Such questions assume that the discipline of history is already in place, with all its interests and standards. Instead he was considering historical inquiry as human activity and how it fits into the larger picture of human existence as a whole. He was addressing the question of why we are interested in the past in the first place, why we should undertake to formulate questions about the past along with methods and procedures for answering them.

Similarly, Husserl's concept of historicity would have certain implications for our understanding of historical knowledge. Though this is not spelled out in the *Crisis,* we could interpret his view as running parallel to what he says about the natural sciences and the life-world. Husserl's argument is that we can only understand the scientific project if we trace its cognitive accomplishments back to their origin in the world of everyday, prescientific experience. Scientific knowledge does not result from the interaction of a priori concepts with passive sense data (this is part of Husserl's criticism of Kant) but arises out of and is directed back to the "always pregiven" life-world. In keeping with its role as background for modern objective natural sciences, the life-world in the *Crisis* is portrayed as a prescientific natural world, the world of perception, perceived things, experienced space-time and causality. For Husserl this is the world we inhabit prior to and independently of the cognitive interests and activities that make up natural science and issue in its particular interpretation of reality.

In the same way, we could say that cognitive interests and activities of our historical disciplines presuppose a broader human and historical life-world, which figures in our ordinary experience, whether we are historians or not. Historical claims and accounts do not emerge ex nihilo from the heads of historians, but presuppose a predisciplinary and pre-objective sense of the past that we share in virtue of our membership in our community. Thus just as the concept of the life-world enables us to understand better the growth and significance of the natural sciences, so the idea of historicity contributes to our understanding of history as a discipline—not by showing how it explains things or by deciding whether it is capable of making objective claims about the past, but by giving us a sense of the larger context and background from which it emerges and differentiates itself.

But for Husserl in the *Crisis*, the epistemological implications of the concept of historicity are not limited to its role in *historical* inquiry. Historicity is an essential feature of all inquiry, including scientific and even mathematical inquiry. What this means is that, for any cognitive project, consciousness does not stand passively before a domain of objects and then undertake on its own a theoretical cognition of that domain. For any given individual, the enterprise of cognition exists as a project before he or she takes it up. The engagement of the individual in such a project presupposes membership in the community and the existence of a tradition of inquiry. In taking up the project, the individual inquirer takes over its questions, goals, concepts, and methods. He also builds on results already obtained by others. Thus a particular science, while it is indeed pursued by individuals, owes its undertaking in each case, as well as its capacity to progress, to the social context in which it exists. There is, of course, a negative side to this: To the extent that research takes the work of its predecessors for granted, it moves further from the original insights that motivated it. Theory can become increasingly abstract or, as Husserl calls it, inauthentic. This can produce the need to return to and reactivate those insights, a need that is harder and harder to meet the further one gets from the original source.

This is the historical and somewhat paradoxical path of inquiry that Husserl describes in the text that has acquired the title *The Origin of Geometry*. While there is a nod to the mythical, "undiscoverable Thales of geometry," and to the problem of understanding the initial jump from the practical mastery of space to its conceptual idealization, the primary subject of this text is how a discipline like geometry, once it is launched, continues on its way, how it exists as a historical continuum or tradition, and how the individual's mastery and eventual contribution to such a discipline depends on the tradition. This is where the geological metaphor of sedimentation comes into play.

While Husserl seeks to exemplify the process of sedimentation by looking in this case at one discipline, what he says applies to all endeavors that come under the heading of Wissenschaft. Ultimately, and most interestingly, of course, it applies to philosophy itself, the one discipline that has always been supposed, somehow or other, to encompass all the others. This is why his concern with historicity finds its primary instantiation, in these texts, in discussing the history of philosophy. He wants to make it clear that he is not just a historian of philosophy, discussing the development of some cultural phenomenon "which might as well be Chinese, in the end"—that is, as observed and described from the outside by some *Geisteswissenschaftler* or anthropologist who is not involved, who is trying to be objective (*Crisis*, § 15). No, the history of philosophy must be approached precisely by those who are engaged in the project, in order to understand the project and the nature of their own engagement in it.

What Husserl realizes at the time of the *Crisis* is that philosophy itself is a community with its own historical background, and to engage in it is to take up a tradition that already exists. Rather than a static collection of eternal questions, it exists as an ongoing inquiry; even if one is motivated to reject current solutions and come up with new ones, one has inherited the questions from the past. The most important philosophers, of course, have been those who come up with new questions rather than new answers; but even they depend on the spiritual inheritance of philosophy. This is the chief implication of the concept of historicity for Husserl, and it is a realization to which he comes rather belatedly. As I have said before, the preoccupation with history is that which distinguishes these late texts most of all from Husserl's earlier work.

His attitude toward the history of philosophy had previously resembled that of his admired model, Descartes. As he wrote in 1910, philosophy had always aspired to be a rigorous science. But so far it had utterly failed. So why waste time with failures? Inquiry should proceed "not from philosophies but from things and from the problems connected with them"—*von den Sachen und Problemen* ("Philosophy as Rigorous Science," p. 146). In *Ideas I*, before he even got around to introducing the phenomenological epoché, he proposed what he called the "philosophical epoché," which means that "we completely abstain from judgment respecting the doctrinal content of all pre-existing philosophy, and conduct all our investigations under this abstention" (*Ideas I*, § 32). Husserl had nothing against the history of philosophy, of course, but like many philosophers before and since, he thought one could draw a clear line between "doing" philosophy and doing its history. Not only that: Phenomenology was originally conceived, I think, as a kind of return to innocence: casting off the prejudices of the philosophical tradition, and even the ultimate prejudice of the natural attitude itself, in order to achieve a pure and unrestricted grasp of experience.

Husserl had always recognized that it is not easy to bracket the natural attitude; hence the laborious attention he pays to refining the phenomenological reduction. Now he has come to recognize that historical prejudices, too, are not easy to overcome. Instead they must be reflected upon and worked through. Now he joins that company of philosophers who believe that philosophy must be done historically if it is to be done responsibly. All theoretical inquiry, even that of the hardest of sciences, is intrinsically historical. This is not usually recognized by those involved; nor should it be, in the case of most disciplines, since the point is to develop theories, not to reflect on them philosophically. But philosophy, unlike other disciplines, is under the obligation to reflect on its own nature as well as that of its subject matter, to try to understand its own procedures even as it practices them.

This, of course, is where the idea of Europe comes in. Reflecting on the historical community of philosophy, Husserl sees it as a European project, tracing its origin to the Greeks. Philosophy has a beginning in a cultural time and place. It is a cultural-historical formation. One of its distinctive features, however, according to Husserl, is its early recognition of the distinction between cultural particularity and universally valid truth. As a particular community, philosophy has always tried to transcend itself and achieve a universal perspective. This paradoxical idea—that of a universal perspective—is really Hegel's idea of the in-itself-for-itself, the ideal of absolute knowledge, which surmounts its own historicity. Husserl does not affirm anything like this, much as he would like to. How could philosophy ever know that it had freed itself from its historical prejudices? He recognizes that this caution leads him into the vicinity of the historical relativism he criticizes in Dilthey. Perhaps philosophy, along with Europe, is in the end nothing more than a particular cultural formation, its universalist aspirations nothing but a quaint—but also sometimes dangerous—feature of its *Weltanschauung* or world-view.

But just as he rejects an absolutist metaphysics, Husserl also refuses to accept the historicist antimetaphysics that incoherently proclaims the impossibility of any transhistorical truth. Phenomenology was never a metaphysics or an antimetaphysics, but something more like a research program, a project. And there it remains for Husserl: as a project of universality that is aware of its own particularity and historicity.

§ 6 Science, History, and Transcendental Subjectivity in Husserl's *Crisis*

Michael Friedman

It is well known that Husserl first makes the "transcendental" (as opposed to merely "empirical") character of his proposed new science of phenomenology completely clear and explicit in "Philosophy as Rigorous Science," published in 1910–11. In particular, although "phenomenology and psychology must stand in close relationship to each other, since both are concerned with consciousness," they nevertheless are concerned with consciousness in two very different ways and from two very different points of view: "Psychology is concerned with 'empirical consciousness,' with consciousness from the empirical point of view, as empirical being in the ensemble of nature, whereas phenomenology is concerned with 'pure' consciousness, i.e., consciousness from the phenomenological point of view" ("Philosophy as Rigorous Science," p. 91). Husserl proceeds to illustrate this contrast, as is also well known, via a comparison with Galileo's mathematization of physical nature in creating an exact natural science. Just as, in physics, properly empirical investigation, on the Galilean model, must be preceded by an a priori mathematical delineation of the essential structure of the domain of inquiry (mathematical essential analysis of the concepts of velocity, acceleration, and so on), so in psychology all properly empirical investigation must be preceded by an a priori phenomenological delineation of the essential structure of the domain of consciousness (phenomenological essential analysis of the structure of perception, recollection, and so on). All merely naturalistic psychology (and thus all psychophysics) is therefore, according to Husserl, in the same insecure position as pre-Galilean science of nature. Only when psychology (like physics) becomes grounded and established on the basis of a preceding a priori—and to this extent transcendental—delineation of its subject matter, can psychology (like physics) become a science, strictly speaking.

These ideas represent the core of Husserl's critique of what he calls *naturalistic philosophy*—the attempt, characteristic of the late nineteenth century (in such thinkers as Helmholtz, Mach, Wundt, and James, for example), to set philosophy on the secure path of a science by closely associating it with recent empirical advances in psychology and psychophysics. Husserl's argument, in opposition to such naturalistic tendencies, is that empirical advances alone can never amount to a truly rigorous science and, in particular, that psychology can become a genuine science only on the basis of phenomenological transcendental philosophy—an investigation "directed towards a scientific essential knowledge of consciousness, towards that which consciousness itself 'is' according to its essence in all its distinguishable forms," an investigation that is analogous to Galileo's mathematical essential analysis of physical nature but that, unlike Galileo's, is also "purely descriptive" and "direct" rather than "exact" and "indirect" ("Philosophy as Rigorous Science," p. 89). Nevertheless, and this is the core of Husserl's critique of what he calls "historicism and *Weltanschauung* or worldview philosophy" (as represented, above all, by Dilthey), phenomenological philosophy is just as much a rigorous science as Galilean mathematical science of nature, insofar as "science is a title standing for absolute, timeless values," and "the 'idea' of science . . . is a supratemporal one, [where] here that means limited by no relatedness to the spirit of one's time" (ibid., p. 136). In particular, the direct or descriptive methods of the new phenomenology, involving an immediate intuitive grasp of essences (*Wesenserfassung*), yield truths that are just as "supratemporal" as the indirect or exact methods of mathematics: "Thus the greatest step our age has to make is to recognize that with philosophical intuition in the correct sense, the phenomenological grasp of essences, a limitless field of work opens out, a science that without all indirectly symbolical and mathematical methods, without the apparatus of premises and conclusions, still attains a plenitude of the most rigorous cognitions, which are decisive for all further philosophy" (ibid., p. 147).

Famously, in his last great work, *The Crisis of European Sciences and Transcendental Phenomenology*, written in the years 1934–37, Husserl returns to the theme of the relationship between phenomenological philosophy and Galilean mathematical science, and to the relationship between transcendental phenomenology and history. Husserl is explicitly responding, in particular, to the circumstance (the present "crisis") that the very idea of science or reason as it has been passed down to us from the Renaissance, through Galileo, and then the Enlightenment—where "reason is a title for 'absolute,' 'eternal,' 'supratemporal,' 'unconditionally' valid ideas and ideals"—has now become profoundly questionable (*Crisis*, § 3). Husserl responds to this crisis of his own time by an essentially historical investigation—a self-consciously idealized, reconstructive, or "teleological" historical investigation—into how the original idea of science that inspired the Renaissance was subsequently obscured or covered over. Specifically, the original ideal of a new " 'philosophical' form of [human] existence: Freely giving oneself, one's whole life, its rule through pure reason or philosophy," resulting in "universal knowledge, absolutely free from prejudice, of the world and man" was subsequently replaced by the ideal of "positive" or "objective science" (as exemplified, above all, by the mathematical and physical sciences); and, most importantly, the reasons for this transformation have themselves remained hidden from us (ibid.). Through an idealized historical reconstruction of this process, Husserl aims to show that transcendental phenomenology is precisely its immanent end, or telos, and that this philosophy alone offers us a satisfactory resolution of our current crisis.

The key figure in the fateful transformation of the idea of science was of course Galileo. For Galileo first realized the Renaissance ideal of "universal [rational] knowledge, absolutely free from prejudice" by creating the mathematically exact science of nature. Here, using reason's own products, the tools and methods of "pure geometry," one was able to secure a science of nature entirely free from all superstition, theology, and metaphysics; and one was then able, accordingly, to implement an ideal of "objective science" or "objective truth" as an infinite, never to be completed task of systematic inquiry into this rationally constituted domain—the domain of mathematically described physical nature. At the same time, however, one thereby lost sight of the fact that mathematics itself (here the domain of pure geometry) is essentially a schematization or idealization of something more fundamental, namely, "the only real world, the one that is actually given through perception, that is ever experienced and experienceable—our everyday life-world [*Lebenswelt*]" (*Crisis*, § 9h). In particular, the actual intuitively given shapes and figures that we can perceive in ordinary life are not perfectly precise or exact but essentially rough and approximate; and their perfectly precise and exact geometrical counterparts arise only on the basis of a further constituent of real or ordinary experience—the "practical art of surveying" or "measurement"—as we imagine purely in thought an idealized indefinite extension of the continual increase in precision actually available empirically (straighter and straighter lines, flatter and flatter planes, and so on). When Galileo and modern mathematical science then turn around and declare that only mathematically described nature is objectively real, and that the "subjective-relative" domain of our actual experience or perception is a misleading appearance of this absolute and objective realm, they forget that our new mathematical description of nature only has sense and meaning on the basis of its necessary origin in the ordinary world of perception and experience—otherwise it is a mere empty formalism. The actual intentional meaning and origin of geometry becomes hidden from us by a historical process of "sedimentation" and "technization," and this leads to the most profound philosophical misunderstandings.

The most fundamental of these misunderstandings is the creation of mind-body dualism by Descartes. Nature as a whole divides into two separate parts: external or extended nature, described solely by pure geometry, and internal or thinking nature, characterized by the essentially nonspatial predicates of thought or consciousness. These two separate parts of nature as a whole are in causal interaction, however, and the merely subjective appearance of external nature that we are actually conscious of in perception is nothing but the effect of purely geometrical matter as it impinges upon our sensory organs. From the Galilean-Cartesian starting point, we thus obtain a naturalized conception of consciousness or the mind as the complement, as it were, of the de-perceptualized, idealized, and mathematized conception of physical nature characteristic of modern science. And the only way to combat such a hopelessly misleading conception of consciousness, according to Husserl, is to recognize that Galilean-Cartesian "physicalist objectivism" itself has a counterpart in modern "transcendental subjectivism." This latter tendency, too, has its origin in Descartes, but it reaches its culmination, in the modern period, in the explicitly transcendental philosophy of Kant—where consciousness or the mind is no longer seen as a complementary part of nature at all, but rather as the transcendental or constitutive ground of all of nature, including, especially, our new mathematical representation of physical or external nature. Nevertheless, Kant's version of transcendental philosophy, for Husserl, was also ultimately a failure; for Kant grounded it on a "transcendental logic" derived from the Leibnizean tradition and thereby misunderstood its essentially intuitive or perceptual dimension. The inevitable result was a "merely regressive" method of transcendental inquiry, whereby we start with the "fact" of mathematical natural science and then develop hypothetical or "mythical" constructions (within Kantian "transcendental psychology") to explain the conditions of possibility of this fact. Just as, in the case of Galileo, the life-world is "the forgotten meaning-fundament of natural science" (*Crisis*, § 9h), "Kant's unexpressed 'presupposition'," according to Husserl, is precisely "the surrounding world of life, taken for granted as valid" (*Crisis*, § 28). It is the task of transcendental phenomenology systematically to "inquire-back [*rückfragen*]" into this "pre-given life-world" so as finally to articulate a fully coherent—and fully scientific—version of transcendental philosophy.

The concept of the pregiven life-world is thus the fundamental concept in Husserl's late reformulation of the task of transcendental phenomenology. But what exactly is this life-world? It is just the world as it is ordinarily experienced or lived by each of us from our own subjective-relative points of view. It is the world now spread out around me in space and time, for example, containing physical bodies simply as they are perceived by me. It is a world in which I myself am embodied and may move, accordingly, among the physical bodies spread out around me in space so as successively to experience various sides or aspects of these bodies from different points of view at different times. In so doing, moreover, I undertake various intentional activities and practical projects, each associated with various ends or values, including such projects as the arts, crafts, and sciences—all with their various "techniques," both practical and theoretical. Finally, it is a world containing other human beings like myself, among whom and with whom I undertake such activities collectively. It is an essentially intersubjective world, in other words, of collective human activity, collective human culture, and collective human history. All of this, as Husserl says, is perfectly obvious to all of us; and the concept of the life-world, so understood, was already present in Husserl's very first articulations of transcendental phenomenology. Thus, in "Philosophy as Rigorous Science," Husserl emphasizes that "experience in the pre-scientific sense . . . plays an important role within the technique proper to natural science" and then states that "the natural sciences have not in a single instance unraveled for us actual reality, the reality in which we live, move, and are" (*Crisis*, § 36). Similarly, in § 47 of *Ideas I*, Husserl contrasts the world of modern mathematical physics with the realm of "experiencing consciousness" or "confirmatory experience [*ausweisende Ehrfahrung*]" on which it is based, and in the following § 48, he emphasizes that this realm of actual experience is intersubjective. It is so far entirely unclear, therefore, what the new systematic status of the life-world within transcendental philosophy is supposed to be. It is similarly unclear what this new status—if indeed it is new—has to do with the new emphasis on philosophical and scientific history also characteristic of the *Crisis*.

The systematic status of the life-world begins to be clarified in the immediately following sections of the *Crisis*, where Husserl discusses "the problem of a science of the life-world" and then uses this discussion to motivate a reconceived version of the "transcendental epoché" or "transcendental reduction" (§§ 37ff.). The problem of a science of the life-world, it turns out, is precisely to distinguish this kind of science from all objective or positive science. The latter arises, as we have suggested, on the basis of the life-world, as we attempt systematically to sort out exactly which entities intuitively appearing in the life-world are "actual," exactly which statements we might make about the life-world are "true." For, in going about our practical affairs within the life-world, it may happen that an entity that originally appeared to be actual (a statement that originally appeared to be true) may turn out not to be what it originally appeared: In walking up to or around an apparent physical object, for example, it may turn out to be a hallucination or a dream. Within the life-world itself, we have inductive regularities and causal relations that guide us in this process of continual correction (real or actual physical objects are perceived to have a backside when we walk around them, for example); and objective or positive science naturally arises from this process as we then attempt, in turn, continually to correct and improve such inductive regularities by reference to more accurate and comprehensive scientific laws. The outcome is an infinite, never-to-be-completed project of systematic inquiry and continual refinement, which is analogous, in this respect, to the indefinite process of continual idealization that first gave rise to the objective science of pure geometry. By contrast, the task of a science of the life-world itself is oriented in precisely the opposite direction. Rather than leaving the immediately intuitive realm of the life-world behind in an infinite process of continual idealization and abstraction, as all properly objective or positive sciences essentially and legitimately do, the point of a science of the life-world is self-consciously to remain entirely within the immediately intuitive realm so as to become fully clear and explicit about the fact that the original intentional meaning of all the properly objective sciences derives solely from their relationship to the life-world.

Thus, whereas all possible objective sciences—mathematics, physics, history, psychology, and so on—must have their own forms of "evidence" or "insight" (their own forms of "essential intuition" and "essential analysis"), these forms of evidence are by no means ultimately or originally intuitive in the sense of the life-world:

> All conceivable verification leads back to these modes of evidence [the "immediate presence" of things as experienced in the life-world] because the "thing itself" (in the particular mode) lies in these intuitions themselves as that which is actually, intersubjectively experienceable and verifiable and is not a substruction of thought; whereas such substruction, insofar as it makes a claim to truth, can have actual truth only by being related back to such evidences. . . . One must fully clarify, i.e., bring to ultimate evidence [*letzten Evidenz*], how all the evidence of objective-logical accomplishments, through which objective theory (thus mathematical and natural-scientific theory) is grounded in respect of form and content, has its hidden sources of grounding in the ultimately accomplishing life, the life in which the evident givenness of the life-world forever has, has attained, and attains anew its prescientific ontic meaning. From objective-logical evidence (mathematical "insight," natural-scientific positive-scientific "insight," as it is being accomplished by the inquiring and grounding mathematician, etc.), the path leads back, here, to the primal evidence [*Urevidenz*] in which the life-world is ever pre-given. (*Crisis*, § 34d)

And it follows, therefore, that "the objective is precisely never experienceable as itself; . . . the experienceability of something objective is no different from that of infinitely distant geometrical structures and, in general, from that of all infinite 'ideas,' including, for example, the infinity of the number series" (ibid.).

This fundamental distinction between a science of the life-world as such and all objective sciences that then may be erected on the basis of the life-world—a distinction that itself arises, as we have seen, from a historical reconstruction of the origin and significance of Galilean natural science for the modern period—is now used to motivate and inform Husserl's reconceived version of the transcendental epoché. Indeed, the very first step on the road to this epoché is what Husserl calls "the epoché of objective science," whereby we turn away from the project of objective science and orient ourselves instead toward the originally intuitive basis, which can alone provide a transcendental or constitutive grounding of these sciences. Yet this first step or epoché is by no means sufficient to define the characteristic attitude or point of view of transcendental phenomenology, for there are two very different ways of making the life-world into the subject matter of an essential or eidetic science: Either we can remain within the life-world itself, in our natural and naive everyday attitude and then attempt, from this point of view, to describe the essential or a priori formal structure of this world (this, in effect, is what we have already done above, when we first introduced the structure of the life-world in terms that are obvious to everyone). Or we can take up "a consistently reflective attitude towards the 'how' of the subjective manner of givenness of the life-world and life-world objects" (*Crisis*, § 38). It is only in this second step that we obtain "the genuine transcendental epoché," and thus the "transcendental reduction," that is, "the discovery and investigation of the transcendental correlation between world and world-consciousness" (*Crisis*, § 41). In this way, in particular, the life-world becomes a "subject matter for a theoretical interest determined by a universal epoché with respect to the actuality of the things in the life-world" (*Crisis*, § 44).

The life-world, to begin with, has a subjective-relative mode of existence. It is given to a particular person from a particular point of view—for example, it is my perceptual world as it is directly given spread out around me now in space and time. In my natural attitude within this world, I have perceptual and theoretical interests in its "truth" or "actuality," and these interests, in turn, are inextricably connected with my practical interests in negotiating my way within this world in pursuit of my various ends and projects. I need to know, for example, whether when I walk up to the wall, I can continue right on through it, and thus my natural attitude within the life-world is essentially bound up with an interest in truth. Indeed, it is for precisely this reason, as we have seen, that the life-world itself necessarily and inevitably leads to the further project of objective science. In order to take up an attitude toward the life-world that decisively and definitively precludes all objective science, so as then to enable a distinctively transcendental science of the life-world qua transcendental ground of objective science, I need to redirect my orientation within the life-world away from all questions of truth or actuality with respect to its objects. I no longer care, for example, whether when I walk up to the wall, I can continue right on through it. Rather, I am concerned solely with those manners of "subjective givenness" of walls (and of all life-world objects more generally) on the basis of which, in the natural attitude, it is then possible to say whether something is one kind of life-world object rather than another (a physical object, say, instead of a hallucination or a dream). For example, the subjective mode of givenness of an actual physical wall, within my experience of the life-world in general, is such that certain visual experiences (of the wall) are correlated with certain kinesthetic experiences (of moving my body up to the wall) and certain further tactual and kinesthetic experiences (of touching the wall, feeling its pressure, and finding the motion of my body inhibited). The transcendental meaning of any object within the life-world (actual physical wall, hallucinated wall, and so on) is thus an intentional correlate, as it were, of some such system of subjective correlations. The life-world itself, as lived in the natural attitude, is a constitutional achievement of transcendental subjectivity; it itself has a transcendental ground—which, since the life-world is always "my" world as it is given to me here and now, is finally "the absolutely unique, ultimately functioning ego" (*Crisis*, § 55).

In the midst of arriving at this new understanding of the transcendental reduction, Husserl inserts an intriguing remark. Immediately after commenting that the reduction "has [now] attained a self-understanding in principle which procures for these insights and for the epoché itself their ultimate meaning and value," Husserl writes:

> I note in passing that the much shorter way to the transcendental epoché in my *Ideas toward a Pure Phenomenology and Phenomenological Philosophy*, which I call the "Cartesian way" (since it is thought of as being attained merely by reflectively engrossing oneself in the Cartesian epoché of the *Meditations* while critically purifying it of Descartes's prejudices and confusions), has a great disadvantage: while it leads to the transcendental ego in one leap, as it were, it brings this ego into view as apparently empty of content, since there can be no preparatory explication; so one is at a loss, at first, to know what has been gained by it, much less how, starting with this, a

> completely new sort of fundamental science, decisive for philosophy, has been attained. Hence also, as the reception of my *Ideas* showed, it is all too easy right at the very beginning to fall back into the naïve-natural attitude—something that is very tempting in any case. (*Crisis*, § 43)

This remark therefore invites us to consider whether there are any fundamental differences between the transcendental (or phenomenological) reduction as it is described in the *Ideas* and as it is now being described in the *Crisis*.

The phenomenological reduction, in the *Ideas*, takes its starting point by "bracketing" or "disconnecting" the entire world experienced in the natural attitude. We do this, of course, by employing (in a modified form) the method of Cartesian "universal doubt," with the following result:

> The entire world, posited [*gesetzte*] in the natural attitude, actually prefound in experience, taken completely "free of theory," as it is actually experienced and clearly exhibited in the interconnections of experiences, now counts as nothing for us—it is to be bracketed, untested but also uncontested. In the same way all positivistic or otherwise grounded theories and sciences that relate to this world, no matter how good, succumb to the same fate. (*Ideas I*, § 32, my translation)

Thus, the *Ideas* begins by simultaneously bracketing, via the method of Cartesian doubt, *both* what Husserl later calls the life-world and all objective sciences that emerge from it, whereas the *Crisis* begins by contrasting the life-world with the objective sciences that emerge from it and then bracketing the latter on behalf of the former.

Of course, as we have seen, this epoché of the objective sciences on behalf of the life-world is only the first step toward the full transcendental epoché developed in the *Crisis*, and Husserl then goes on to take the more radical step of leaving the natural attitude for the sake of "a consistently reflective attitude towards the 'how' of the subjective manner of givenness of the life-world and life-world objects" involving "a universal epoché with respect to the actuality of the things in the life-world." And there is no doubt, in particular, that this second, more radical step bears a close similarity to the phenomenological epoché of the *Ideas*, resulting in the new phenomenological "region of pure consciousness." Nevertheless, there is also a fundamental dissimilarity. The *Ideas* arrives at the realm of pure consciousness by a principled distinction between "immanent" and "transcendent perceptions" (or "acts"), where the former are such "*that their intentional objects, when these exist at all, belong to the same stream of experience as they do,*" and the latter are not. Thus my awareness of my own conscious states counts as an immanent perception, and my awareness of things in the world around me is transcendent (*Ideas I*, § 38). The crucial point, as Husserl then explains, is that transcendent perceptions are always dubitable, whereas immanent perceptions are indubitable. The phenomenological region of pure or "absolute" consciousness thereby emerges as the "residuum after the nullifying of the [transcendent] world"—it emerges as that which is absolutely secure from all doubt (*Ideas I*, § 49).

In the *Crisis*, by contrast, there is no use whatsoever of the method of Cartesian doubt. In particular, transcendental subjectivity and the transcendental ego are not arrived at by "nullifying" the life-world to which they are attached. Rather, transcendental subjectivity and the transcendental ego are arrived at by changing our orientation within the life-world itself so that its own transcendental structure—its own intentional constitution—is clearly and explicitly revealed. To be sure, in the course of this process of reorientation, we must indeed bracket all questions of actuality or truth pertaining to the things of the life-world. But the transcendental epoché now involves no properly skeptical questions at all, and there is absolutely no discussion, in particular, about the dubitability of "outer," or transcendent, experience as opposed to the indubitability of "inner," or immanent, experience. As we have seen, all questions of actuality and truth are bracketed here simply to preclude, at the outset, any move away from the life-world toward objective science, and to focus our attention instead on the ultimate transcendental conditions that make possible both the life-world itself, as it is taken in the natural attitude, and the objective sciences that necessarily and inevitably arise from it.

In the final section of Part III A of the *Crisis*, Husserl sums up his new understanding of the transcendental epoché and the corresponding type of "apodicticity" appropriate to transcendental phenomenology in an especially striking way:

> From this one also understands the sense of the demand for apodicticity in regard to the ego and all transcendental knowledge gained upon this transcendental basis. Having arrived at the ego, one becomes aware of standing within a sphere of evidence of such a nature that any attempt to inquire-

> back behind it would be absurd. By contrast, every ordinary appeal to evidence, insofar as it was supposed to cut off further inquiry-back, was theoretically no better than an appeal to an oracle through which a god reveals himself. All natural evidences, those of all objective sciences (not excluding those of formal logic and mathematics), belong to the realm of what is "obvious," what in truth has a background of incomprehensibility. Every evidence is the title of a problem, with the sole exception of phenomenological evidence, after it has reflectively clarified itself and shown itself to be ultimate evidence. It is naturally a ludicrous, although unfortunately common misunderstanding, to seek to attack transcendental phenomenology as "Cartesianism," as if its *ego cogito* were a premise or set of premises from which the rest of knowledge (whereby one naively speaks only of objective knowledge) was to be deduced, absolutely "secured." The point is not to secure objectivity but to understand it. One must finally achieve the insight that no objective science, no matter how exact, explains or ever can explain anything in a serious sense . . . . The only true way to explain is to make transcendentally understandable. (*Crisis*, § 55)

Thus, whereas the exposition of the transcendental epoché in the *Ideas* could (and did) easily lead to the prevalent misunderstanding of transcendental phenomenology as a form of epistemological foundationalism, aimed at finally securing our objective knowledge in a basis of absolute certainty, the exposition in the *Crisis* is self-consciously intended to remove this misunderstanding once and for all.

We can deepen our appreciation of this point, and we can connect it, in turn, with the problem of historicity, if we now take a brief look at Part III B of the *Crisis*. Here Husserl takes up the thread of the (reconstructive) historical narrative interrupted by Part III A, beginning with "Kant's unexpressed 'presupposition': the surrounding world of life, taken for granted as valid," and continues the story into the nineteenth century. His focus is on the profoundly problematic relationship between psychology and transcendental philosophy during this period, and thus on the very problematic ("naturalistic philosophy") which first prompted Husserl to clarify the peculiarly transcendental (as opposed to merely empirical) character of the new science of phenomenology in "Philosophy as Rigorous Science." Now, in the *Crisis*, the main theme of "the philosophical development after Kant" is characterized as a "struggle between physicalistic objectivism and the constantly reemerging 'transcendental motif'" (*Crisis*, § 56). Physicalistic objectivism, as we already know, obtains a naturalized conception of the mind, consciousness, or (as Husserl now puts it) the soul as the complement, within nature as a whole, of the de-perceptualized, idealized, and mathematized conception of physical nature characteristic of post-Galilean modern science. Nature as a whole is thus "seen 'naturalistically' as a world with two strata of real facts regulated by causal laws . . . [and] souls [are] seen as real annexes of their physical living bodies (these being conceived in terms of exact natural science) . . ." (*Crisis*, § 61). We obtain the familiar dualistic conception (which Husserl characterizes as "absurd") according to which "body and soul thus signify two real strata in this experiential world [that is, the life-world] which are integrally and really connected similarly to, and in the same sense as, two pieces of a body. Thus, concretely, one is external to the other, is distinct from it, and is merely related to it in a regulated way" (*Crisis*, § 62). And this conception is absurd, for Husserl, because "it is contrary to what is essentially proper to bodies and souls as actually given in life-world experience, which is what determines the genuine sense of all scientific concepts" (*Crisis*, § 62). In particular, in the life-world itself, I am directly and immediately given both physical bodies spread out around me in space and my own physical body in which I "hold sway"—as essentially *em*-bodied—as I live and move among them.

Nevertheless, it then appears, at least at first, that in order to delimit the subject matter of a proposed new objective science—a new science of "descriptive psychology"—it is still possible to bracket all physical or bodily dimensions of the life-world by means of what Husserl calls "a fully consciously practiced method . . . which I call the *phenomenological-psychological reduction*—taken in this context as a method for psychology" (*Crisis*, § 69). I arrive at this standpoint, namely, by an "epoché of validity" with respect to all beliefs concerning perceptual physical objects in the life-world around me, so that "purely descriptive psychology thematizes persons in the pure internal attitude of the epoché, and this gives it its subject matter, the soul" (ibid.). In other words: "For the psychologist, as long as he limits himself to pure description, the only simple objects are ego-subjects and what can be experienced 'in' these ego-subjects themselves (and then only through the epoché) as what is immanently their own, in order to make this the subject matter of further scientific work" (*Crisis*, § 70).

But precisely here Husserl now finds a fundamental problem. We have been attempting to find a foundation for "individual psychology"—by "performing the universal reduction in such a way that it is exercised individually upon all individual subjects accessible through experience and induction, and in each case" (*Crisis*, § 71). We have been assuming as "obvious" that "human beings are external to one another, they are separated realities, and so their psychic interiors are also separated. Internal psychology can thus be only individual psychology of individual souls, and everything else is a matter for psychophysical research" (ibid.). However, "the properly understood epoché, with its properly understood universality, totally changes all the notions that one could ever have of the task of psychology, and it reveals everything that was just put forward as obvious to be a naiveté which necessarily and forever becomes impossible as soon as the epoché and the reduction are actually, and in their full sense, understood and carried out" (ibid.).

The fundamental problem, it turns out, is that the present version of the epoché—the phenomenological-psychological reduction—entirely misconstrues the a priori constitutive force of *inter*subjectivity:

> How, more precisely, does each [subject] have world-consciousness while it has self-apperception as this human being? Here we soon see, as another a priori, that self-consciousness and consciousness of others are inseparable; it is unthinkable, and not merely [contrary to] fact, that I be a human being in a world without being *a* human being. There need be no one in my perceptual field, but fellow humans are necessary as actual, as known, and as an open horizon of those I might possibly meet. Factually I am within an interhuman present and within an open horizon of humankind: I know myself to be factually within a generative framework, in the unitary flow of an historical development in which this present is humankind's present, and the world of which it is conscious is an historical present with an historical past and an historical future. (*Crisis*, § 71)

After the properly understood universal epoché, then: "What remains, now, is not a multiplicity of separated souls, each reduced to its pure interiority, but rather, just as there is a single universal nature as a self-enclosed framework of unity, so there is a single psychic framework, a total framework of all souls, which are united not externally but internally, namely, through the intentional interpenetration which is the communalization of their lives" (ibid.).

The phenomenological-psychological epoché is thereby transformed, in particular, into the truly universal *transcendental* epoché:

> This means at the same time that, within the vitally flowing intentionality in which the life of an ego-subject consists, every other ego is already intentionally implied in advance, by way of empathy and the empathy-horizon. Within the universal epoché which actually understands itself, it becomes evident that there is no separation of mutual externality at all for souls in their own essential nature. What is a mutual externality for the natural-mundane attitude of world-life prior to the epoché, because of the localization of souls in living bodies, is transformed in the epoché into a pure, intentional, mutual internality. With this the world—the straightforward existing world and, within it, existing nature—is transformed into the all-communal phenomenon "world," "world for all actual and possible subjects," none of which can escape the intentional implication according to which it belongs in advance within the horizon of every other subject.
>
> Thus we see with surprise, I think, that in the pure development of the idea of a descriptive psychology, which seeks to bring to expression what is essentially proper to souls, there necessarily occurs a transformation of the phenomenological-psychological epoché and reduction into the *transcendental*; and we see that we have done and could do nothing else here but repeat in basic outlines the considerations that we had to carry out earlier in quite another interest, i.e., in the interest not of a psychology as a positive science but of a universal and then transcendental philosophy. (*Crisis*, § 71)

In particular, if I undertake the transcendental epoché as it is articulated in the *Crisis*, I arrive at a position of necessary intersubjectivity or human community, according to which "all souls make up a single unity of intentionality within the reciprocal implication of the life-fluxes of the individual subjects, a unity that can be unfolded systematically by phenomenology" (ibid.).

We noted above that Husserl had already emphasized the importance of intersubjectivity in § 48 of the *Ideas*. In particular, he there suggests that "the factually separate worlds of experience [of different subjects] fit together through the interconnections of actual experience into a single intersubjective world, the correlate of the unitary spiritual world (the universal extension of the human community)" (*Ideas I*, § 48). But this theme, in the *Ideas*, is only thus briefly suggested; and, most importantly, there is no discussion at all of the necessary historicity of the community in question—of the circumstance that, as described in the *Crisis*, "factually I am within an inter-human present and within an open horizon of humankind: I know myself to be factually within a generative framework, in the unitary flow of an historical development in which this present is humankind's present, and the world of which it is conscious is an historical present with an historical past and an historical future" (*Crisis*, § 71).

If we take this last idea seriously, however, it follows that the old dream of "Philosophy as Rigorous Science"—the idea of pure phenomenology as a fundamentally ahistorical discipline concerned only with "'absolute,' 'eternal,' 'supratemporal,' 'unconditionally' valid ideas and ideals"—must also be given up. The role of transcendental phenomenology, in this respect, is completely disanalogous to the role of Galilean pure geometry in grounding, for the first time, a mathematical science of nature. Rather, the role of transcendental phenomenology, as Husserl now understands it, is precisely to intervene in our historically present (circa 1935) "crisis of European sciences" so as to recapture the original idea of science as a new "'philosophical' form of [human] existence: freely giving oneself, one's whole life, its rule through pure reason or philosophy," resulting in "universal knowledge, absolutely free from prejudice, of the world and man" (*Crisis*, § 3). In particular, all objective or positive science is now seen to be an achievement of transcendental subjectivity, depending (among other things) on "the essential structures of absolute historicity, namely, those of a transcendental community of subjects as one which, living in community through intentionality in these most general and also in particularized a priori forms, has in itself and continues to create the world as intentional validity-correlate, in ever new forms and strata of a cultural world" (*Crisis*, § 72). Objective science is no mere "technique" for cognitively accessing (and controlling) an alien nonhuman reality; human subjectivity is no mere "residuum" of a de-perceptualized external nature. On the contrary, the world of objective science is a free, rational creation of historical human subjectivity, now seen as an equally free and rational historical transcendental community. Transcendental phenomenology thereby restores the intrinsic worth and dignity of both objective science and that human subjectivity—*our* human subjectivity—of which it is a necessary product.

§ 7 Universality and Spatial Form

Rodolphe Gasché

According to Husserl, "the *teleological beginning*, the true birth of the European spirit as such" (*Crisis*, § 15), lies in the Greek primal establishment of the idea of a universal science whose foundation in intersubjectively reconstructable truths makes it a science that in absolute self-responsibility accounts for all of its claims. Although the European spirit is born in Greece, it manifests itself properly only with another primal establishment that occurs in the Renaissance, and "which is at once a reestablishment [*Nachstiftung*] and a modification of the Greek primal establishment" (*Crisis*, § 15). But the reestablishment of the idea of an all-encompassing rational science at the beginning of modernity is not a wholesale underwriting of the Greek heritage. It is a modification that transforms the idea in question. This reestablishment is responsible not only for the modern sciences' indisputable accomplishments but also for the current crisis of the sciences, that is, for what Husserl diagnoses as the sciences' inability to account for the meaning of their own activity, and, hence, their loss of any relation to humanity's basic concerns. Perhaps more importantly, the success and simultaneous crisis of the European sciences is also an indication of the problematic nature (if not the narrowness) of the concept of universality that informs the modern sciences. Indeed, as Husserl remarks, "a definite [*bestimmtes*] ideal of a universal philosophy and its method forms the beginning; this is, so to speak, the primal establishment of the philosophical modern age and all its lines of development" (*Crisis*, § 5). Hereafter, I will discuss in some detail the founding event of the modern sciences in the Renaissance, primarily in order to elicit what, precisely, this specific concept of universality amounts to and what its intrinsic limitations are in comparison to the Greek idea of universality. The aim of this elucidation is to show that, for Husserl, the concept of universality that dominates the modern sciences fails to make good on the promise of a universal horizon, which emerged as a task with Greek philosophy. Furthermore, since Husserl links the European spirit to this promise, one must conclude that, essentially, the modern sciences have also failed that spirit.

The project of an all-encompassing science, or metaphysics, which originates in ancient Greece, is not only a science of the *one* world—the world that encompasses all the relative worlds. It is one that, in order to secure access to this one world and to establish for it terms that are in principle intelligible to all, independently of their race, gender, customs, culture, religion, nationality, and so forth, presupposes an attitude that is critical of everything that is of the order of such particulars. This universal science embodies the ideal of a community freeing itself precisely from all traditions, and traditionalisms, and shaping itself freely according to insights of reason that are recognizable for their universality. However, the primal establishment of the new philosophy that characterizes the Renaissance, and which, according to Husserl, coincides with "the primal establishment of modern European humanity itself—humanity which seeks to renew itself radically, as against the foregoing medieval and ancient age, precisely and only through its new philosophy" (*Crisis*, § 5), lacked this critical attitude with respect to the Greek heritage. The Renaissance takes it over as "*an unquestioned tradition*" (*Crisis*, § 9g), and it thus essentially misses what is so essential about the Greek project. Indeed, by its very nature the project and the task of a humanity that understands itself from the *one* world cannot be taken over slavishly. As the *Crisis* suggests, the ancient model of a universal rational science, which implies the critical rejection of tradition, precludes being appropriated in a traditionalist spirit. Husserl writes that the ancient model "was not to be taken over blindly from the tradition but must grow out of independent inquiry and criticism" (*Crisis*, § 3). The very spirit of the Greek conception of reason demanded a critical attitude toward the model in question as a heritage bequeathed upon Europe as well as a free and independent reactivation of this heritage.[1] But if the very idea of a universal and rational science radically excludes all uncritical acceptance of any received heritage (including that of itself), then it is also the case, as Husserl notes, that the way in which ancient philosophy, in its first, original establishment seeks to realize the universal task of philosophy is not without its own naiveties and inherent limits.[2] In fact, although this task is, as we have seen before, a remarkably strange one, Greek philosophy soon lost sight of its strangeness in the very attempt to develop an all-encompassing science. Furthermore, by conceiving of this task as self-evident, "the naive obviousness of this task [became] increasingly transformed . . . into unintelligibility," with the result that "reason itself and its [object], 'that which is,' became more and more enigmatic" (*Crisis*, § 5). If reason has become enigmatic, it is, Husserl suggests, because the wonders that mathematics and physics have accomplished, in particular, "the wonderful symbolic arts of the 'logical' construction of their truths and theories," have become "incomprehensible" to the extent that the reason or meaning for their existence is no longer evident.[3] As Husserl suggests, the Renaissance philosophers did not question what had been handed down from antiquity, and therefore, "the first invention of the new idea [of a universal science in the Renaissance] and its method allowed elements of obscurity to flow into its meaning" (*Crisis*, § 9g). These elements prevented the exact sciences from achieving "knowledge about the *world*," that is, about and for the world shared by all. Foregoing any reflective inquiry into the original meaning of the received conceptions, the new sciences adopted the universal insights of Greek geometry and mathematics without questioning their origin and proceeded to develop a kind of disengaged universality on the basis of the mathematization of nature and the formalization of mathematics. The abstraction and emptiness of the latter—although not altogether illegitimate, and above all, highly successful—not only became increasingly severed from the concerns of humanity as such, but also remained tied to one particular world, one particular mankind, and one particular horizon—to Europe as a particular *ethnia*. The reestablishment and modification of the Greek idea of a universal rational science by the modern sciences thus obfuscates the true spirit of the primal establishment of the idea of a universal science in Greece. It follows from this that "the spectacle of the Europeanization of all other civilizations" that begins with the Renaissance, rather than bearing witness "to the rule of an absolute meaning, one which is proper to the sense . . . of the world," may be, in Husserl's own words, for the time being at least, "a historical non-sense" (*Crisis*, § 6). Indeed, what is exported under the guise of the techno-sciences is a kind of universality that no longer has any relation to the *one* world, the one in which we all live.

If the modern age, an age characterized by its rediscovery of philosophy as a universal task, is "not merely a fragment of the greater historical phenomenon" constituted by the inaugural establishment of philosophy in Greece (*Crisis*, § 5), it is because this rediscovery is not a simple repetition of that event. Husserl writes that

> as the reestablishment [*Neustiftung*] of philosophy with a new universal task and at the same time with the sense of a renaissance of ancient philosophy—it is at once a repetition and a universal transformation of meaning. In this it feels called to initiate a new age, completely sure of its idea of philosophy and its true method, and also certain of having overcome all previous naiveties, and thus all skepticism, through the radicalism of its new beginning. (*Crisis*, § 5)

Indeed, the reestablishment of philosophy as universal rational science during the Renaissance marks a radically new beginning, in that the modern age reshapes the universal task at the heart of philosophy, and thus also reshapes the very meaning of universality. But this reformulation of the universal task comes with its own naiveties, which are a function of the way the Renaissance relates to the Greek heritage.

Philosophy as "universal science, science of the universe, of the all-encompassing unity of all that is," which, according to *The Vienna Lecture*, arises in Greece as a result of the "*new sort of attitude* of individuals toward their surrounding world" (*The Vienna Lecture*, p. 276)—that is, a radically critical attitude—is the science of the *one* world, the world for everybody, that is, of humanity itself. The discovery of this *one* world enables the conception of a universal science in Greece—a science concerned with what is universal, and which seeks to proceed according to principles and rules that are reconstructable by everyone. Now the science that sustained this idea of universality in Greece, and that served to flesh out the conception of the *one* intersubjectively shared world, is first and foremost geometry. Its pure forms, its ideal shapes of space-time, which are constructed according to rules that are verifiable at all times, and which permit everyone to reproduce them identically, have absolute and universal value with regard to the one and same world shared by everyone. Husserl writes:

> Out of the undetermined universal form of the life-world, space and time, and the manifold of empirical intuitable shapes that can be imagined into it, [geometry] made for the first time an objective world in the true sense—i.e., an infinite totality of ideal objects which are determinable univocally, methodically, and quite universally for everyone. Thus mathematics showed for the first time that an infinity of objects that are subjectively relative and are thought only in a vague, general representation is, through an a priori all-encompassing method, objectively determinable and can actually be thought as determined in itself or, more exactly, as an infinity which is determined, decided in advance, in itself, in respect to all its objects and their properties and relations." (*Crisis*, § 9b)

As Husserl observes, "scientific acquisitions, . . . after their method of assured successful production has been attained . . . are imperishable; repeated production . . . produces in any number of persons something identically the same, identical in sense and validity" (*The Vienna Lecture*, pp. 277–78). What is thus found to be identical, and to obtain for all relative worlds—the pure forms of space and time—are imperishable idealities. Indeed, "what is acquired through scientific activity is not something real but something ideal," which itself, moreover, becomes "material for the production of idealities on a higher level, and so on again and again" (ibid.). This discovery of pure geometric shapes is not that of "mere spatiotemporal shapes" abstracted from bodies experienced in the intuitively given surrounding world. Nor are they arbitrarily imagined shapes, or shapes transformed by fantasy (*Crisis*, § 9a). In distinction to the imaginary (hence, still sensible) idealities of pure morphological types, such as roundness, for example, which have never the perfection that allows for their absolute identical repetition, the pure forms of geometry—such as the circle—are, as Husserl writes, "limit-shapes," that is, identical and invariant idealities obtained by way of a passage to the limit (*Crisis*, § 9b).[4] These idealities arise from "a peculiar sort of mental accomplishment" (*Crisis*, *Appendix V*, p. 348), which Husserl terms an "idealizing accomplishment," and thus possess "a rigorous identity" (*Crisis*, *Appendix II*, p. 313), which submits to "the conception of the 'again and again' . . . *in infinitum*," a repetition that in nature differs from the open endlessness characteristic of abstract figures.[5] "The great invention of idealization" (*Crisis*, § 9h) by geometry and mathematics, which provides "the pure shapes it can construct *idealiter* [in the form of an idea]" (*Crisis*, § 9), and in such a manner that anyone can reconstruct them, is what permits an insight of universal scope into the *one* objective world shared independently of all particularities.[6]

The universal science recast in the Renaissance, along with its distinctly new conception of universality, thus rests on a rediscovery of ancient geometry. It is a discovery, however, that dispenses with the task of reconstructing what had given birth to it in ancient Greece. The latter's accomplishments are taken for granted. As Husserl submits, geometrical methodology, which permits overcoming "the relativity of subjective interpretation" and attaining "something that truly is"—"an identical, nonrelative truth of which everyone who can understand and use this method can convince himself"—is a given for Galileo, and he takes it over "with the sort of naiveté of a priori self-evidence that keeps every normal geometrical project in motion" (*Crisis*, § 9b). Furthermore, this identical truth revealed by geometry is understood as the truth of nature. In fact, for him, "everything which pure geometry, and in general the mathematics of the pure form of space-time, teaches us, with the self-evidence of absolute, universal validity, about the pure shapes it can construct *idealiter*," belongs to "true nature" (*Crisis*, § 9). It must also be said that the geometry that Galileo inherited was "a relatively advanced geometry," one that had already become "a means for technology, a guide in conceiving and carrying out the task of systematically constructing a methodology of measurement for objectively determining shapes in constantly increasing 'approximation' to geometrical ideals, the limit-shapes" (*Crisis*, § 9b). Its proven effectiveness was one more reason to exempt it from the need for questioning. According to the *Crisis*, Galileo received this heritage in such a way that "he, quite understandably, did not feel the need to go into the manner in which the accomplishment of idealization originally arose (i.e., how it grew on the underlying basis of the pre-geometrical, sensible world and its practical arts) or to occupy himself with questions about the origins of apodictic, mathematical self-evidence" (*Crisis*, § 9b).[7] For Galileo, the original strangeness of geometrical idealization, that is, the discovery of pure shapes and their universally reconstructable evidence, is gone. Geometry is for him an unquestioned cultural acquisition. Without having to reflect back on its genesis, the self-enclosed world of its pure forms can be manipulated like any other cultural tool. Geometry's evidences have become self-evident; in other words, rather than reflecting on the origin of geometry—a reflection that would have permitted Galileo to link it to the universal and transcendental eidetic structures of the pre-geometrical life-world—he takes it over uncritically, as an abstract truth. It did not dawn on Galileo to make geometry, "as a branch of universal knowledge of what is (philosophy), [and] geometrical self-evidence—the 'how' of its origin—into a problem" (ibid.). However, Galileo in turn develops a conception of universal science that, even though it has become obvious for us today, was definitely *merkwürdig*, strange, or odd at the time. The notion of universality characteristic of the modern sciences that comes into being with Galileo is a strange notion as well. But its oddity is distinct from that which characterizes the universal in the primal establishment of the European spirit in Greece. The strangeness of the universality peculiar to the modern sciences derives, as we will see, from its alienation from the life-world. The strangeness of this new conception of universality is that of the merely abstract.[8] At any rate, since the Renaissance slavishly takes the truths of geometry as abstract givens, the radically novel conception of universality of the emerging natural sciences (that marks the inception of modernity), notwithstanding the fact that it will have been a clear advance over the Greek notion of a universal science, is tinged with naïveté. As Husserl holds, this naïveté is responsible for the current crisis of the sciences.

Of what, then, does the new idea of the universality of the sciences consist? According to Husserl, the sciences inherited from the ancients—"Euclidean geometry, and the rest of Greek mathematics, and then Greek natural science"—undergo "an immense change of meaning" during the Renaissance. This change, which primarily affects mathematics, namely geometry and the formal-abstract theory of numbers and magnitudes, sets new tasks for its disciplines, "tasks of a style which was *new in principle*, unknown to the ancients," in being not only "*universal* tasks," but infinite tasks (*Crisis*, § 8). Indeed, in spite of the Greeks' idealization of empirical numbers, units of measurement, and empirical figures in space; in spite of their transformation in geometry of propositions and proofs into ideal-geometrical propositions and proofs; and, finally, in spite of their understanding of Euclidian geometry as "a totality of pure rationality, a totality whose unconditioned truth is available to insight and which consists exclusively of unconditioned truths recognized through immediate and mediate insights," Husserl nonetheless holds that "Euclidian geometry, and ancient mathematics in general, knows only finite tasks, a finitely closed a priori" (ibid.). The Greek world is a finite world—a cosmos within natural limits whose unexceedable horizon encloses all the *pragmata* of mortal beings. The discovery of infinite ideals, and hence infinite tasks, is, for Husserl, a positive achievement of modernity, and represents a clear advantage that the modern sciences have over the ancient ones. Antiquity, Husserl notes in the *Crisis*, does not grasp "the possibility of the infinite task which, for us, is linked as a matter of course with the concept of geometrical space and with the concept of geometry as science belonging to it" (ibid.). As Jacques Derrida has argued, as a passage to the limit, geometric idealization is by definition the infinite transgression of the sensibly ideal morphological shapes of the life-world. Hence, the inaugural idealization that opened Greek geometry endows it from the outset with infinite fecundity. Yet this infinitization "no less *first* limits the a priori system of the productivity. The very content of an infinite production will be confined within an a priori system which, for the Greeks, will always be *closed*" (Derrida, 1978, 127).[9] In distinction from the Greeks, for whom the ideal and universal knowledge of geometry was limited to a finite number of forms or shapes for which it furnished a rational foundation, the moderns take geometry as a science capable of accounting for all possible forms. Husserl writes: "To ideal space belongs, for us, a universal, systematically coherent a priori, an infinite, and yet—in spite of its infinity—self-enclosed, coherent, systematic theory which, proceeding from axiomatic concepts and propositions, permits the deductively univocal construction of any conceivable shape which can be drawn in space" (*Crisis*, § 8). The very notion of ideal space to which the Greeks arrived by idealizing empirical figures contains, for us, in ideal form, all possible spatial shapes. For us, as opposed to the Greek finite understanding, ideal space is "a rational infinite totality of being," "an infinite world" of idealities whose ideal objects "become accessible to our knowledge [not] singly, imperfectly, and as it were accidentally, but as one which is attained by a rational, systematically coherent method. In the infinite progression of this method, every object is ultimately attained according to its full being-in-itself" (ibid.). With "the actual discovery and conquest of the infinite mathematical horizon," one that is not only limited to ideal space but is soon extended to numbers as well, the task of the sciences becomes infinite, one of infinite universal tasks. However, the radicality of this reconfiguration of the concept of universality does not come to a close with mathematics. Indeed, as Husserl remarks, the latter's rationalism "soon overtakes natural science and creates for it the completely new idea of *mathematical natural science*—Galilean science" (ibid.). Posing "the radical *problem of the historical possibility of 'objective' science*, objectively scientific philosophy," it is not merely a matter, for Husserl, "of establishing science's historical, factual point of origin in terms of place, time, and actual circumstances, of tracing philosophy back to its founders, to the ancient physicists, to Ionia, etc.; rather, it must be understood through its original spiritual motives, i.e., in its most original *meaningfulness* [*Sinnhaftigkeit*] and in the original forward development of its meaningfulness" (*Crisis, Appendix V*, p. 347). Only by following Husserl through his discussion of Galileo's mathematization of nature, and the genesis of modern natural sciences, will we be able to evaluate the full extent of the reformulation of the concept of universality in the Renaissance and its underlying unquestioned self-evidences.[10]

Even though the idea of the mathematization of nature is something that today is taken for granted, it was initially a rather strange idea, as Husserl contends. In my analysis of the underpinnings of this idea, I will highlight this strangeness, not only because it is, as I have argued elsewhere, a constitutive aspect of universality (one that has drawn little attention as far as I am able to judge), but also because the kind of strangeness that Husserl associates with the universality of the modern sciences differs in kind from the one associated with the Greek idea of universality.[11] It is perhaps not insignificant that while speaking of the Greek project of a universal rational science, Husserl uses the German term *merkwürdig*, remarkable, but also strange, or odd, whereas the modern sciences are described as a *befremdliche Konzeption*, that is, as a conception that appears strange, if not displeasing or disconcerting (*Crisis*, § 9c). Husserl begins his discussion of Galilean science by noting that for the Greeks, ideality, identity, and universality are primarily characteristics of the realm of the pure forms. For them "the real has [only] a more or less perfect methexis in the ideal" (*Crisis*, § 9). Husserl therefore concludes that all application of the pure forms of geometry to nature was only "a primitive application" (ibid.). Now with Galileo, nature no longer participates in a realm of idealities distinct from it; it itself is idealized and is shown to possess its own ideal substratum. Husserl writes that "through Galileo's *mathematization of nature, nature itself* is idealized under the guidance of the new mathematics; nature itself becomes—to express it in a modern way—a mathematical manifold" (ibid.). Needless to say, with this idealization of a domain initially foreign to the ideality of pure forms, nature becomes the object of universal insights. But what is it that allows pure mathematics and geometry to become the guide to the formation of exact physics in the first place? What is, Husserl asks, the "hidden, presupposed meaning" of Galileo's guiding model of mathematics, which "had to enter into his physics along with everything else," that is, with everything that "consciously motivated him" (*Crisis*, § 9a). If I am interested in Husserl's answer to this question, it is precisely because it will tell us something significant about the idealities and the universality peculiar to the modern sciences.

Even though the abstraction of mere spatio-temporal shapes from things intuited in the world of everyday life never on its own leads to the formation of geometrically ideal, that is, absolutely identical shapes, the perfection of technical capabilities in the practical world allows for the experience of the progressive precision of these shapes, and hence of "an open horizon of *conceivable* improvement" regarding these shapes (ibid.). As Husserl remarks, "out of [this] praxis of perfecting, of freely pressing toward the horizons of *conceivable* perfecting 'again' and 'again,' *limit-shapes* [*Limes-Gestalten*] emerge toward which the particular series of perfecting tend, as toward invariant and never attainable poles" (ibid.). From the repetitive attempts aimed at perfecting, say, the measurements of shapes, the ideal limit-shapes, or pure shapes, which subsequently become the object of geometry, arise by way of acts of idealization, that is, by way of acts that Husserl characterizes as distinctly different from acts of abstraction. Unlike the shapes abstracted from spatio-temporal bodies, these limit-shapes are pure idealities, that is identically repeatable forms that are "intersubjectively determinable, and communicable in [their] determinations, for everyone" (ibid.). Husserl writes: "If we are interested in these ideal shapes [for their sake] and are consistently engaged in determining them and in constructing new ones out of those already determined, we are 'geometers.' The same is true of the broader sphere which includes the dimension of time; we are mathematicians of the 'pure' shapes whose universal form is the co-idealized form of space-time" (ibid.). Next to real practice, geometry and mathematics (whose objects are the pure limit-shapes, or spatio-temporal forms, that emerge in real praxis where they can only be infinitely approximated) thus give rise to "an *ideal* praxis of 'pure thinking' which remains exclusively within the realm of pure limit-shapes" (ibid.). However, what also sets this ideal praxis radically apart from empirical praxis is the fact that in pure mathematics and geometry, these shapes are no longer the object of graduation and approximation. In mathematical praxis exactness is attained "for there is the possibility of determining the ideal shapes in absolute identity, of recognizing them as substrates of absolutely identical and methodologically, univocally determinable qualities" (ibid.). But besides the idealizations of all sensibly intuitable shapes (such as straight lines, triangles, and circles) that mathematics and geometry can carry out according "to an everywhere similar method," it also becomes possible to use "these elementary shapes, singled out in advance as universally available, and according to universal operations which can be carried out with them, to *construct* not only more and more shapes which, because of the method which produces them, are intersubjectively and univocally determined" (ibid.). Indeed, the discovery that characterizes modern geometry was that of the possibility of "producing constructively and univocally, through an a priori, all-encompassing systematic method, *all* possible *conceivable* ideal shapes," (ibid.) whether or not there are sensibly intuitable models for them in reality. With this, the ideal space has become infinite as has the task, "which for us, is linked as a matter of course with the concept of geometrical space" (*Crisis*, § 8).

Yet, as Husserl notes, although geometry construes the entirety of all conceivable shapes in thought alone—and seemingly, in complete abstraction from the practical world, geometrical methodology "points back to the methodology of determination by surveying and measuring in general, practiced first primitively and then as an art in the prescientific, intuitively given surrounding world" (*Crisis*, § 9a). Since the shapes that are intuitively experienced, or that are merely conceived in general (through abstraction), blend into one another in the "open infinity" of the space-time continuum of the prescientific surrounding world, they are "without 'objectivity.'" They are not "intersubjectively determinable, and communicable in [their] determinations, for everyone—for every other one who does not at the same time factually see" them (ibid.). Now the role of the art of measuring already consists in securing some intersubjective objectivity for these shapes. It serves to render univocally determinable each single shape that is experienced in the space-time continuum of shapes characteristic of everyday life, and this according to a methodology that is intersubjectively grounded. By "picking out as [standard] measures certain empirical basic shapes, concretely fixed on empirically rigid bodies which are in fact generally available," and holding these against other bodies, the art of measuring becomes capable of determining the latter "intersubjectively and in practice univocally—at first within narrow spheres (as in the art of surveying land), then in new spheres where shape is involved [*Gestaltsphären*]" (ibid.). What Husserl seeks to bring into the open here is the ultimate rootedness of pure mathematics in the life-world. The purely geometrical way of thinking, and hence, the striving for philosophical knowledge, that is, "knowledge which determines the 'true,' the objective being of the world," is the idealization of "the empirical art of measuring and its empirically, practically objectivizing function." As is made clear in the "The Origin of Geometry," the art of measuring is "pregiven to the philosopher who did not yet know geometry but who should be conceivable as its inventor" (*Origin of Geometry*, p. 376). Even though the "philosopher proceeding from the practical, finite surrounding world . . . to the theoretical world-view and world-knowledge . . . has the finitely known and unknown spaces and times as finite elements within the horizon of an open infinity," still he does not yet have "geometrical space, mathematical time, and whatever else is to become a novel spiritual product out of these finite elements which serve as material; and with his manifold finite shapes in their space-time he does not yet have geometrical shapes" (ibid.). Undoubtedly, the philosopher becomes the proto-geometrician only on the basis of a new sort of praxis, one that arises from pure thinking, but also one that takes its clues from the praxis of the gradual perfection of the art of measuring. For Husserl, then, the art of measuring is clearly "the trailblazer for the ultimately universal geometry and its 'world' of pure limit-shapes." Pure mathematics and geometry have their origin in this method for securing intersubjective truth, and it is this origin that provides them with their true meaning. This is the premise on whose basis Husserl argues that, by taking the achievements of these disciplines for granted, Galileo had become oblivious to geometry's and mathematics' origin in that life-world, an origin that alone makes them meaningful for mankind. But something else becomes clear at this juncture as well, namely that the rediscovery of ancient geometry in the Renaissance amounted not only to having recourse to "a tradition empty of meaning" (*Origin of Geometry*, p. 366) but also that the prime (if not the sole) way of seeking to secure universal intersubjective validity in modern Europe takes place by way of spatio-temporal shapes and forms. Yet Husserl's prime concern is to demonstrate that the new sciences that have come into being by modeling themselves after ancient geometry are disconnected from the prescientific life in the given world that represents the horizon of all meaningful inductions.[12] But by highlighting only Galileo's obliviousness to geometry's origin in the life-world and the ensuing consequences for the development of the modern sciences, the fact that universality is primarily ascribed to geometrical idealities, and that this priority of the universality of spatio-temporal shapes affects the very concept of universality itself, may fail to receive in Husserl's work the attention it merits. Undoubtedly, geometry is the first philosophical science because it permits the establishment of the absolute identity of ideal shapes in such a way that they are the same for everyone at any time. But the possibility of universality thus becomes, first and foremost, a function of the shape of the *res extensa*, of things of nature, more precisely of their idealized shapes, and this to such a degree that in the absence of such shapes it seems to be impossible to secure anything universal at all. The fate of "psychology," which in the wake of the Cartesian dualism of nature and mind has never been capable of achieving the status of a science comparable to that of the natural sciences, is a clear indication of the limitation of the modern concept of universality to the universality of the idealized shapes of spatio-temporal bodies.

In *The Origin of Geometry*, Husserl argues that the persisting truth-meaning of geometry is a function of the possibility of producing invariant and apodictically general contents for the spatio-temporal sphere of shapes, contents that can be idealized and can "be understood for all future time and by all coming generations of men and thus be capable of being handed down and reproduced with the identical intersubjective meaning." But universality is not equivalent to idealized spatio-temporal shape, as the privilege accorded, from the Greeks via the Renaissance to the present, to what is extended in the world of bodies would seem to suggest. When Husserl adds, "This condition is valid far beyond geometry for all spiritual structures which are to be unconditionally and generally capable of being handed down," he does not wish to imply that universality is predicated on ideal spatio-temporal shapes alone, but rather that in order to achieve an intersubjectively recognizable universality, "the apodictically general content [alone of other than geometrical forms], invariant throughout all conceivable variation" (*Origin of Geometry*, p. 377), is to be taken into account in the idealization. Indeed, the possibility of idealization and universality is not exclusively linked to geometrical form; hence the concept of universality itself is not intrinsically limited to shape in the spatio-temporal sense. As *The Origin of Geometry* points out, "'ideal' objectivity . . . is proper to a whole class of spiritual products of the cultural world, to which not only scientific constructions and the sciences themselves belong but also, for example, the constructions of fine literature" (*Origin of Geometry*, pp. 356–57). Language (that is, language in general) is particularly a domain from within which "ideal objects" and "ideal cognitive structures" arise (*Origin of Geometry*, pp. 357, 364). In spite of the preeminence that the ideal shapes of geometry have enjoyed in the sciences, and in a conception of philosophy *more geometrico*, that is, as Husserl points out, a conception that embraces the "methodological ideal of physicalism," they constitute only one of the possible formations capable of ideal objectivity and of "intersubjective being" (*Crisis*, § 34f). Yet the exemplarity of geometrical universal truth is not therefore diminished. Indeed, as Derrida has shown, a distinction made by Husserl in *Experience and Judgment* bears on these different types of ideality. Compared to the idealities of words or cultural products that are *bound idealities*—in that they are dependent on an empirically determined temporality or factuality—the geometrical ideal objectivities are *free idealities* (though free only with respect to empirical subjectivity), and thus the only ones that can claim to be truly universal (Derrida, 1978, 71–72).

Still, the intricate connection between spatio-temporal shapes and rational universal ideality will have to continue to interest us as we now turn to Galileo's mathematization of nature. In fact, as we will see, the very possibility of the modern exact sciences rests on this connection. Even though pure geometry and mathematics reveal identical and nonrelative truths, these truths pertain only to bodies in the world—to the bodily world. According to the *Crisis*, Galileo realized that all "*pure* mathematics has to do [solely] with bodies and the bodily world only through abstraction, i.e., it has to do only with *abstract shapes* within space-time, and these, furthermore, as purely 'ideal' limit-shapes" (*Crisis*, § 9b). Galileo was thus fully aware of the fact that the universal truth of which pure mathematics is capable concerns exclusively an abstraction of the bodily world, its abstract shapes, and, in the end, only its ideal and fully identifiable limit-shapes. Pure mathematics derives its universal truths from the bodily world alone, more precisely from the idealized shapes of these very bodies. Yet there is much more in the physical world than just bodies. Or, differently put, the spatiality of bodies is only one of the eidetic components of bodies. Hence if there is to be a philosophical or scientific knowledge of the world, the specific qualities, or sensible plena that all actual shapes possess in empirical sense-intuition, as well as "the universal causal style" by which all experienced bodies are bound, must necessarily be accounted for. Needless to say, it is mathematics once again that shows Galileo the way to accomplish this task. The latter had not only shown that through the idealization of subjectively relative objects, one can arrive at objectively determinable entities, but also that by descending again from the world of idealities to the empirically intuited world, as demonstrated by the contact between mathematics and the art of measuring, "one can universally obtain objectively true knowledge of a completely new sort about the things of the intuitively actual world, in respect to that aspect of them (which all things necessarily share) which alone interests the mathematics of shapes, i.e., a [type of] knowledge related in an approximating fashion to its own idealities" (ibid.). Indeed, by becoming "applied geometry," ideal geometry made it possible for the art of measuring to calculate, for everything in the world of bodies, "with compelling necessity, on the basis of given and measured events involving shapes, events which are unknown and were never accessible to direct measurement" (ibid.). Galileo thus concluded that it should be possible to do for all the other aspects of nature—the real properties and the real-causal relations of bodies in the intuitable world—what had been done for the sphere of shapes, namely, to extend "the method of measuring through approximations and constructive determinations" (ibid.), that is, a method developed exclusively with respect to shapes, to their altogether different realm. However, a difficulty arises at this point: How can a science or philosophy of the one and same world that binds us all be achieved if "the material plena—the 'specific' sense-qualities—which concretely fill out the spatio-temporal shape-aspects of the world of bodies *cannot*, in their own gradations, be *directly* treated as are the shapes themselves" (ibid.)? Exactitude is possible only with respect to idealities. Even though sensible qualities are subject to gradation, in their case, as well as in the case of everything that is of the order of the concrete sensibly intuited world, it is, as Husserl writes, "difficult for us to carry out the abstract isolation of the plena . . . through a universal abstraction opposed [*in universaler Gegenabstraktion*] to the one which gives rise to the universal world of shapes" (*Crisis*, § 9c). As a result, no precise measurement of them is possible nor "any growth of exactness or of the methods of measurements" (ibid.). If no direct mathematization of the plena is possible, it is because there seems to be no world of idealities specifically their own. In short there are no limit-plena, and hence also no "geometry" of such idealities. As Husserl quite unambiguously remarks: "We have not two but only *one* universal form of the world; not two but only *one geometry*, i.e., one of shapes, without having a second for plena" (ibid.). What this means is that with respect to the objective world, or nature, we possess, as Jan Patočka formulates it, "only one rational and general form to whose ideal objectivity no parallel in the domain of quality exists" (Patočka, 1988, 233–34). As far as the one and the same world is understood objectively, that is, as a bodily world, there is only one form of universality, and this form of universality is inherently thought from this one aspect that all bodies have in common, namely shape.

Motivated by the Greek idea of an all-embracing science, Galileo concluded that in order to account for the world of nature as the one and the same objective world that we all share, those aspects of it that (unlike the shapes of bodies) cannot directly be mathematized, and which, because they lack a mathematizable world-form, are heterogeneous to spatio-temporal forms, can nevertheless be mathematized, although in an oblique way. Indeed, since in every application to intuitively given nature, pure mathematics must renounce its abstraction from the intuited plena, without therefore having to give up what is idealized in the shapes, Galileo realized that "in one respect this involved the performance of co-idealization of the sensible plena belonging to shape" (*Crisis*, § 9d), and that, consequently, the intuited plena are capable of indirect mathematization. In short, the objective world as a whole "becomes attainable for our objective knowledge when those aspects which, like sensible qualities, are abstracted away in the pure mathematics of spatio-temporal form and its possible particular shapes, and are not themselves directly mathematizable, nevertheless become mathematizable *indirectly*" (*Crisis*, § 9c). Now the indirect mathematization of that part of the world that has no mathematizable world-form is possible only if one assumes that the plena and the shapes of the bodies to which they belong are intertwined. Husserl notes that indirect mathematization is "thinkable only in the sense that the specifically sensible qualities ('plena') that can be experienced in the intuited bodies are closely related [*verschwistert*, that is, like brother and sister] in a quite peculiar and regulated way with the shapes that belong essentially to them" (ibid.).[13] According to this idea, or rather hypothesis, which founds Galileo's new physics, "every change of the specific qualities of intuited bodies which is experienced or is conceivable in actual or possible experience refers causally to occurrences in the abstract shape-substratum of the world, i.e., that every such change has, so to speak, a counterpart in the realm of shapes in such a way that any total change in the whole plenum has its causal counterpart in the sphere of shapes" (ibid.). Even though this conception has lost "its strangeness [*Befremdlichkeit*] for us and [has taken] on—thanks to our earlier scientific schooling—the character of something taken for granted," (ibid.) this was not yet the case for Galileo. We must, Husserl writes, "make clear to ourselves the *strangeness* of his basic conception in the situation of his time" (ibid.). If this idea with which the groundwork was laid for an all-encompassing science of the objective world is strange, it is not merely because of its novelty. Since it is a conception that in the meantime has become obvious to us, and is universally accepted, we can assume that its strangeness derives, first, from the fact that it permits the establishment of something that is universally valid, and that therefore collides with held beliefs. Undoubtedly, the Renaissance had already opened itself to the general idea (which announces itself in everyday experience) that all occurrences in the intuitive world yield to universal induction. But the assumption that "everything which manifests itself as real through the specific sense-qualities must have its *mathematical index* in events belonging to the sphere of shapes" (ibid.) and that makes it possible to indirectly mathematize the plena—that is, the construction *ex datis* (from the facts), and full determination, of all events in the realm of the plena—is also strange in the sense that it is, and always remains, a mere hypothesis. More precisely, this assumption is remarkable and strange because, in spite of its verification in numerous instances, it remains a hypothesis that must endlessly be confirmed. Husserl writes: "The Galilean idea is a *hypothesis*, and a very remarkable one at that [*von einer höchst merkwürdigen Art*]; and the actual natural science throughout the centuries of its verification is a correspondingly remarkable sort of verification. It is remarkable because the hypothesis, in spite of the verification, continues to be and is always a hypothesis; its verification (the only kind conceivable for it) is an endless course of verifications" (*Crisis*, § 9e). The idea that sustains Galileo's physics—the mathematical approach to nature and the universality that it establishes with respect to the plena and causality—remains forever hypothetical. Indeed, since the mathematization of the plena is based on a substruction in thought of a hypothetic relation between shapes and the qualities of spatio-temporal things, that is, on something that can never be experienced and verified as presenting itself as such, and hence in full self-evidence, the verification of such a relation needs to be repeated again and again. The truth of a connection between bodily shapes and plena cannot be acquired once and for all, and, consequently, is never a given.[14] What is strange about Galileo's founding hypothesis, and what puts it at odds not only with the natural attitude but also with the scientific spirit of Galileo's time, is that the universality that it establishes with respect to the one and same world of physics presupposes an infinite task. In spite of the repeated verification of this hypothesis in the praxis of the sciences, it must be continuously reasserted. Undoubtedly, this essential instability of the universal laws of the sensible qualities of intuitable things (and of the overall causal style of these things) derives from the indirect mathematization in which the plena are tied to ideal shapes, that is, essentially, to an order foreign to them. In conclusion, we can say that the infinite need to verify the hypothesis that supports the indirect mathematization of nature, by which the one and same physical world is rendered scientifically and universally intelligible, shows that the infinite task-character of the objective universal derives from this universal's intrinsic foreignness to its object. As Husserl emphasizes, the constant necessity to verify exact physics' founding hypothesis is not caused by possible error but because "in the total idea of physics as well as the idea of pure mathematics [there] is embedded the *in infinitum*, [as] the permanent form of that peculiar inductivity which first brought geometry into the historical world" (ibid.). It is rooted in the distinct foreignness of idealities predicated on spatio-temporal shapes—that is, idealities resulting from the objectification of "one abstract aspect of the world": the pure shapes of "ideal geometry, estranged from the world [*weltentfremdete*]" (*Crisis*, § 9b)—to what even within the objective world is not of the order of the bodily. Furthermore, even though in the indirect mathematization of the plena, "one always has to do with what is individual and factual," the whole method has from the outset, "a *general* sense." Husserl writes: "From the very beginning, for example, one is not concerned with the free fall of *this* body; the individual fact is rather an *example*" (*Crisis*, § 9d).

The factual success and inductive productivity of the approach in question is not in doubt. It infinitely surpasses the accomplishments of all everyday forms of prediction. Nevertheless this success, which remains meaningless as long as its method is not tied back to the concerns of the life-world, is also a function of an intrinsic foreignness of a universal derived from one aspect of the world of bodies to other aspects of the latter. Even though such foreignness is an intrinsic feature of universality as such, the particular way in which the idealized bodily shapes are brought to bear on the sensible qualities of the things of nature (those permitting of sensible experience), that is, the need to confirm infinitely the hypothesis of a link between spatio-temporal shape and sensible plena, indicates an intrinsic limit of this very concept of universality in its application to nonphysical, or non-thinglike, aspects of the world. Let us remind ourselves that, for Husserl, the idea of a universal philosophy, or science, that announces itself in Greece, one that is synonymous with what the name *Europe* stands for, is the idea of an all-embracing philosophy, or science. Such a philosophy cannot limit itself to an intersubjectively binding understanding of the objective world, the world of nature, or to a concept of universality that is only binding for the physical world. Furthermore, given that geometry has been the model for the sciences, the question arises as to the extent to which universality is linked to shape in the first place. In order to achieve an intersubjective consensus about the ideal objectivities of the spiritual products of the cultural world, to which Husserl calls attention in *The Origin of Geometry*, it thus becomes necessary to uncouple universality from idealized shape and to think the form of other ideal objectivities in terms that are no longer tributary to this feature of bodily things.[15]

As a result of the absence of a reflection back on the original meaning-giving achievement of the idealization of the spatio-temporal forms that gave rise to the geometrical ideal constructions, it appeared that geometry produced "a self-sufficient, absolute truth which, as such—'obviously'—could be applied without further ado" (*Crisis*, § 9h). Rather than being understood as "indices of 'inductive' lawfulness of the actual givens of experience," the mathematical limes-formations arrived at through idealization were taken to correspond to the objectively true being of nature.[16] As early as Galileo, a "surreptitious substitution [*Unterschiebung*] [took place] of the mathematically substructed world of idealities for the only real world, the one that is actually given through perception, that is ever experienced and experienceable—our everyday life-world" (Crisis, § 9h). Husserl submits that this substitution of idealized nature for the prescientifically intuited nature is of the order of a disguising or covering over and replacement of the life-world. He writes:

> In geometrical and natural-scientific mathematization, in the open infinity of possible experiences, we measure the life-world—the world constantly given to us as actual in our concrete world-life—for a well-fitting *garb of ideas* [*Ideenkleid*], that of the so-called objectively scientific truths . . . Mathematics and mathematical science, as a garb of ideas, or the garb of symbols of the symbolic mathematical theories, encompasses everything which, for scientists and the educated generally, *represents* [*vertritt*] the life-world, *dresses it up* [*verkleidet*] as "objectively actual and true" nature. It is through this garb of ideas that we take for *true being* what is actually a *method*—a method which is designed for the purpose of progressively improving, *in infinitum*, through "scientific" predictions, those rough predictions which are the only ones originally possible within the sphere of what is actually experienced and experienceable in the life-world. (Ibid.)

If this is the case, if indeed the pregiven world provides the horizon within which and in relation to which the idealization of nature takes place, then the substitution of the objective world of nature for the life-world amounts to an ethico-philosophical error. By reflecting back on what in the life-world motivated the creation of geometry—and, by extension, the sciences that from the Renaissance modeled themselves after it—the accomplishments of geometry and the sciences are not only tied to purposes "which necessarily [lie] *in* this pre-scientific life and [are] related to its life-world" (ibid.); but the idealized limit-shapes, in short, the spatio-temporal universals, also reveal themselves to be the products of acts of concrete intentional consciousnesses. In other words, by bringing into relief the life-world from which all idealizations and intersubjective identifications emerge, the geometrical universal exposes its historicity—that is, its production by a constituting consciousness. With this, the intersubjective accomplishments of geometric idealization, which have given rise to the success of the European sciences, are shown to be the product of a transcendental ego whose accomplishments are the very object of the new *episteme* of phenomenological philosophy sketched out in Part III of the *Crisis*, and which understands itself as the critical renewal of the Greek idea of an all-embracing science.

§ 8 Husserl, History, and Consciousness

Eva-Maria Engelen

Among Husserl's aims in the *Crisis*, one is particularly urgent, namely his search for the origins of meaning. These origins are at the same time the origins for thinking, understanding, and for knowledge.[1] He uses two very similar notions in German for *origin*. One is *Ursprung* and the other one is *Urstiftung*. *Ursprung* is used in the context of evidence and justification, whereas *Urstiftung* appears in the context of datable events. Husserl very often runs these two together.[2] There is a tension between them that runs through his whole discussion of the notion of origin or original evidence and experience. The tension I am speaking of is the one between historicity and ahistoricity (of meaning). Because of this longing for original experience, we have to try, according to Husserl, to go back to historical experiences in order to get back to the original meaning and the original experience. Another source for original experience is consciousness. One aim of this chapter is to show to what extent consciousness as an origin for original experience is also subject to the same tension between historicity and ahistoricity. The other aims are to show how far this tension runs through all of the other sources for original experience and meaning he is speaking of.

There are three philosophically important origins for meaning for Husserl and one personal one. The philosophically relevant ones are (1) consciousness, (2) the life-world, and (3) European philosophy and the history of the sciences. The fourth and rather private origin of meaning concerns the crisis of Husserl's life in the 1930s in Germany and is related to the meaning of one's own life. This crisis started when the Nazis came into power. This rather private topic is (4) his own origin as a Jewish person and a German thinker.

Corresponding to these three or four origins are three,[3] or respectively, four ways or "methods" to arrive at origins.

1. Consciousness: By the epoché, the bracketing of everything that belongs to the world, we can get to the absolute ego and to true subjectivity, according to Husserl.
2. The life-world: By "going back" to the life-world, we can arrive at a deobjectification of scientific thinking and therefore at true meaning.
3. European philosophy and the sciences: By reflection on the history of Greek philosophy and the history of the sciences in Europe, we can again get to a deobjectification of positivistic scientific thinking. One way to do this is by reflection on the sedimentation of philosophical and scientific concepts.
4. His own origin as a (converted) Jewish person and a German thinker: Husserl himself has his origin as a philosopher and thinker in the German nation. And he tries to preserve this origin and therefore his identity by writing the *Crisis*.

It is not only the word origin that makes this comparison plausible. There are deeper, and as I think, more meaningful parallels to the other three topics, as I will show later on.

The tension between historicity and ahistoricity (of meaning) is more or less evident in all of the philosophically important origins but one, namely that of absolute subjectivity or consciousness. In spite of this, I will argue that all three of the philosophically relevant origins mentioned are subject to this tension. Before turning to this point later on, I shall merely hint at the line of argument: Concerning the notion of absolute consciousness, one can show that it is linked to historicity because, even if one conceded that an epoché was possible, and that it could reveal an ahistorical origin of evidence (which one may not be obliged to do), one would still have to posit a concept of historicity as part of the concept of consciousness. Therefore all Husserl can get to is historically valid origins of evidence and meaning. Before discussing this point at length, I want to examine how far Husserl's private crisis and the concern for his own personal origin are linked to the three philosophical origins mentioned above. We shall see that the parallels between the philosophically relevant origins and the concern for his personal origin are quite revealing.

There exists a letter addressed by Husserl in 1936 to the Austrian philosopher Gustav Albrecht in which he reflects on being a member of the German nation and its blood (!) through his thinking and writing: "And it is a difficult thing at our age to find a possible mode of existence once the rug has been pulled out from under one's feet. It requires much spiritual energy, and in order to deal with it, I must counter it with a powerful and superior force of philosophical concentration—thus this extreme struggle in the composition of the last work"[4] (namely, the *Crisis*). Further on he writes: "I have finally at least to make clear for myself that I am no stranger in German philosophy (and therefore in this nation) and that all of the great thinkers of the past, whom I admired so much, and whose thoughts have grown in mine into new kinds, would have had to rank me as a true heir of their spirit, as blood of their blood."[5]

We can find the same tension between historical approach, sedimentation, and ahistoricity in this quotation as we find in his characterization of Greek philosophy as the origin of philosophy and in his description of the life-world. He sees the historical dimension of his own origin as a philosopher as inhering in his being an heir and a continuer of past German philosophy. But in mentioning the German blood that he shares by continuing this heritage, he is employing an ahistorical, biological concept. I do not think that Husserl became a victim of Nazi ideology by doing so. I rather think that he uses this term as a metaphor in order to show that he is not only part of a historical development (in being an heir and continuer of German philosophy) but that he is a *German* philosopher. He does this for his own sake. And he does it because he wants to safeguard his own identity, in order to avoid becoming a mere object for the supposed representatives of the nation he belongs to. His own crisis of identity is due to the historical development in Germany, as is his effort to demonstrate that he is a German philosopher by writing the *Crisis*.

Husserl thereby wants to prove that he is still part of the German nation by writing the *Crisis*, but as a philosopher he also wants to show that he is part of the European philosophical spirit—a spirit that was founded by the Greeks and that was, as Hegel claimed, continued by German philosophy. European philosophy is one of the origins of rationality for

Husserl. Therefore it is also a foundation for a responsible critique of contemporary scientific developments:

> For we are what we are as functionaries of modern philosophical humanity; we are heirs and cobearers of the direction of the will which pervades this humanity; we have become this through a primal establishment [*Urstiftung*] which is at once a reestablishment (*Nachstiftung*) and a modification of the Greek primal establishment [*Urstiftung*]. In the latter lies the *teleological beginning*, the true birth of the European spirit as such.
>
> This manner of clarifying [*Aufklärung*] history by inquiring back into the primal establishment [*Urstiftung*] of the goals which bind together the chain of future generations, insofar as these goals live on in sedimented forms yet can be reawakened again and again and, in their new vitality, be criticized; this manner of inquiring back into the ways in which surviving goals repeatedly bring with them ever new attempts to reach new goals . . . this, I say, is nothing other than the philosopher's genuine self-reflection on what he is *truly seeking*, on what is in him as a will coming *from* the will and *as* the will of his spiritual forefathers. It is to make vital again, in its concealed historical meaning, the sedimented conceptual system, which, as taken for granted, serves as the ground of his private and non-historical work. It is to carry forward, through his own self-reflection, the self-reflection of his own forebears and thus not only to reawaken the chain of thinkers, the social interrelation of their thinking, the community of their thought, and transform it into a living present for us but, on the basis of the *total unity* thus made present, to carry out a *responsible critique*. (*Crisis*, § 15)

The European philosopher is described as a member of a spiritual community that has evolved over time. He is dependent on the concepts he has inherited, but in reflecting on their true historical sense, he is able to criticize contemporary developments. According to Husserl, this activity is the true work of the philosopher. Again, one can see that Husserl is describing the work of a philosopher in the same way in which he writes about his own efforts in the letter to Gustav Albrecht cited above. And one can also say that criticizing as a task of philosophy is also one of the main tasks of the *Crisis*. And as criticizing is part of the project of enlightenment, one can also say that the *Crisis* itself is an attempt at enlightenment, which is to be arrived at by examining origins and thereby finding a neutral standpoint from which one is able to criticize the intellectual developments of one's own times. But criticizing intellectual developments is only one part of the project of enlightenment. Since Kant, self-reflection is the other one. This second way of enlightenment is what Husserl calls in the *Crisis* the true self-reflection of a philosopher in succession of his spiritual predecessors. But as he mentions in the letter to Albrecht, the *Crisis* itself is an attempt at personal self-reflection for Husserl and at finding his own new ground.

Husserl does not only think that sedimented forms of thinking can be reawakened and criticized, but that this enterprise is a most important part of the philosopher's self-reflection. This is because in digging up the sedimented forms and conceptual systems, the philosopher is finding a ground of his private and nonhistorical work. Therefore he is carrying forward his own self-reflection through the self-reflection of his own forebears, as Husserl says. This is at once a reflection on why and how a philosopher should work in a historical perspective and a finding out of a ground for (self-)reflection on his own situation as a philosopher in his specific historical situation.

One way of considering Husserl's thinking here is to compare it to a similar line of argument in the work of Michel Foucault. Husserl, one might say, is deliberately conflating the two notions of Aufklärung that Foucault tries to distinguish in a number of his essays that attempt to answer the Kantian question "Qu'est-ce que la critique?" ("What is enlightenment?") anew.[6] Foucault points out that Kant was the first to ask this question in a historical-critical dimension, in other words in one that is not linked to the epoch of modernity, as it would be if one asks it only in the context of a so-called "philosophy of consciousness." The historical-critical dimension consists in asking for the *historical conditions* of the development of rationality, as opposed to merely inquiring into its timeless nature. On this new conception, enlightenment (Aufklärung) is not primarily a topic for the first person, for the creature who is able to say "I am," for it concerns historical conditions that lead to criticism, and that antedate the thinker posing the question. Asked in this later manner, our question is freed of its dependence on the particular historical epoch in which it is asked, at least to the extent that the origin of the question is now correctly situated in the historical sequence that gave rise to it, and in the sense that we may ask for the *historical conditions* of the development of any rationality.

The "running together" of the two notions of Aufklärung by Husserl is quite subtle, because he invokes a historical prototype for the philosopher who starts from self-reflection as the source for criticism. At the beginning of the *Crisis*, we find a short paragraph where Husserl presents his undertaking in the *Crisis* as similar to Socrates' approach to philosophy:

> I seek not to instruct but only to lead, to point out and describe what I see. I claim no other right than that of speaking according to my best lights, principally before myself but in the same manner also before others, as one who has lived in all its seriousness the fate of a philosophical existence. (*Crisis*, § 7)

This could well be a description of Socrates' way of philosophizing (leading, not instructing, only pointing out . . .) and, in applying this description to himself, Husserl situates himself in the ongoing history of European philosophy and implicitly puts himself at a new beginning of this tradition. We can very well interpret this move as an attempt to give philosophy a new start all while relating it to a historically evolved prototype of the philosopher—one that, as a prototype, is paradoxically ahistorical.

But to identify self-reflection and responsible critique as the main tasks of European philosophy is as much an arbitrary act as is his sublimation of the history of European philosophy. Husserl justifies his method by hinting at the exceptional historical position of Greek thinking. But this supposed exceptionality and uniqueness can only be proven by showing that Greek, and therefore European, philosophy is not one historical datum among others, but that it is exceptional in its being nonhistorical, a paradigm for all time. Husserl touches on other, non-European ways of philosophizing, but only gives hints about why he considers them to be less important, and in so doing, he is following Hegel's *History of Philosophy*.

A further element of ahistoricity or nonhistoricity can be seen in what he calls the "sedimentation of concepts." A sedimented conceptual system is said to serve as the ground of one's private, nonhistorical work, and this despite its concealed historical meaning. This is so because the concealed historical meaning is thought to be an original one and therefore one that is truer. The tension between historicity and nonhistoricity is obvious: One has to look for the most basic historical grounds in order to find nonhistorical ones for one's own systematic work.

I turn now to point three from my list above: European philosophy and the sciences. According to Husserl, a reflection on the history of Greek philosophy and the history of the sciences in Europe permits us to de-objectify or, better, to de-idealize positivist scientific thinking. One way to do this is by reflecting on the sedimentation of philosophical (and scientific) concepts in such a way as to lead us to the original meaning of the concepts. Sedimented concepts and thoughts include the original meanings that are passed on over time, but that are not always immediately at our disposal. For we have to reflect on them. Their historical sedimentation explains why we should be able to grasp them in principle and why we are still linked to them—why they can still mean something to us.

This is a process that Husserl takes for granted in the history of European philosophical thinking and in the history of the sciences. In using concepts and thoughts that transport a sedimented meaning, we participate in the original evidence that they transport. Reflecting on this sedimented meaning and its evidence is a way of self-reflection for Husserl and therefore (again) a philosophical task of particular importance. Once we have access to the sedimented (original) evidence, we are also able to criticize ongoing thinking and work in the contemporary sciences. The sedimented truth gives us a standpoint from which we are able to criticize our own time and to reflect on it—again the most relevant work for a philosopher, especially one who sees himself as standing in a Kantian tradition.

But why does Husserl need the concept of the life-world in addition to the concept of scientific sedimentation? This is an important question, for the life-world need not be *our* life-world. It might well be a historical one. In the case of the sciences, Husserl therefore assumes two forms of historical access to an original meaning.

If the life-world is a historical one, then it is the source that allows us to detect the original, and now sedimented evidence. We cannot have access to sedimented meaning without going back to the former life-world. In Husserl's terms, the mathematician does not possess the actual meaning of the mathematical method he works with, nor of the "implications of meaning which are closed off by sedimentation or traditionalization, i.e., of the constant presuppositions of his [own] constructions, concepts, propositions, theories" (*Crisis*, § 9h). But the mathematician can make them self-evident again, just as he can reactivate the so-called self-evidence by reactivating its meaning. This is achieved by reflecting on the life-world, which Husserl calls the forgotten meaning-fundament of the natural sciences. Let me touch briefly on this point.

In the *Crisis*, a work whose concept of a life-world differs from his other works in that it is the natural world,[7] Husserl argues that returning to the life-world will also give access to original evidence. But might he not be mistaken? For life-worlds are themselves already forms of lives in which we find historically evolved and developed concepts, practices, and planning. Thus they cannot be conceived as entirely free of historicity. This becomes evident when we examine Husserl's example of geometry.

> The geometry of idealities was preceded by the practical art of surveying, which knew nothing of idealities. Yet such a pre-geometrical achievement was a meaning-fundament for geometry, a fundament for the great invention of the ideal world of geometry. . . . Thus it could appear that geometry with its own immediately evident a priori "intuition" and the thinking, which operates with it, produces a self-sufficient, absolute truth, which, as such . . . could be applied without further ado. (*Crisis*, § 9h)

According to Husserl, to regress from geometry back to the life-world it arose from would mean to reflect on the art of surveying. But what he calls the art of surveying might also be called a science. Historical research on the Egyptian art of surveying doesn't provide a justification for why we should not call this science. In fact, historians do call it a science nowadays. And it is equally well a way of idealizing the world and not merely an "applied" science.[8] This example indicates why there is no "science-free" life-world, no world in which we can find an original meaning for the sciences, if we mean by that one that is not already scientifically formed.

Husserl's claim is that in reflecting on the art of surveying, we can see that mathematics and the sciences did not arise merely because they involve a priori knowledge just waiting to reveal itself to the human mind. They were rather developed to help us meet some special need. And I think this is what Husserl also wants us to keep in mind, that is to say the sense in which history of the sciences is supposed to help our understanding. The problems that drove the development of mathematics and the sciences might change over time. Therefore it would not necessarily serve a practical purpose to remind a mathematician nowadays of the purpose geometry once had. But it might be a meritorious feature of good history of science that it reveals how a science was once helpful. Keeping in mind that it took its sense from being useful might make us reflect on the manner in which the contemporary sciences add something useful to people's lives.

Husserl's search for the grounds of original evidence is therefore at least twofold. One side is his reflection on sedimented philosophical and scientific conceptual systems. And the other one is his turning to the life-world—what is supposed to be the empirical, natural world in the *Crisis*. This empirical world is the world of simple causalities, as Husserl says, the one to which one must have access if one is to be able to bracket the scientific theories that order our expectations.[9]

But even if I am right to conclude that there is, in fact, no access to those grounds of original evidence, one still might hold that it is useful to reflect on the original problem a science was meant to help with. Such a reflection might, for instance, be motivated by our identifying, as does Husserl, a crisis or general discomfort with scientific development and its impact on our lives and our self-understanding. Think, for example, of modern medicine, biology, or neurobiology. If we reflect on the philosophical implications medicine had as a science in its beginning, we might see contemporary medicine in a very different light. The philosophical notion of a good life was one that defined the aims of a physician and his science. And the notion of a good life also includes the notion of autonomy and self-determination. This is not the place to dilate on this subject. And there is no doubt that Husserl wanted more than that. He wanted to get back to original *evidences*. But still, we might think it useful to get to a standpoint that enables us to criticize the contemporary sciences by doing history of the sciences—even if this standpoint is not an absolute one.

We ought to keep in mind that the reflection on a life-world is described differently from the bracketing of the epoché, which I will try to describe in the following. What I mean is that Husserl tries to find the original evidence of scientific work by going back to the life-world, but by "bracketing" the life-world in the case of the epoché, he tries to get to the original meaning of subjectivity and consciousness. It is important to recognize that, in the *Crisis*, Husserl tries both to arrive at origins via abstraction—namely the epoché—and that he is at the same time criticizing mathematics and physics for being too abstract, and thereby objectifying life. As we shall see, the epoché is described as a way of bracketing, and I think it is not too far-fetched to claim that this is a metaphor Husserl has taken from logic and mathematics.

In conclusion, I will return to the topic I announced at the beginning of this chapter. How ahistorical or absolute can the concept of consciousness be in the *Crisis*? David Woodruff Smith describes consciousness and nature—the mental and the physical side of experience—as two sides of the same event (Smith, 1995).[10] Our neuroscience, he suggests, presupposes our everyday life-world understanding of ourselves, which is indispensable in the practice of science. But experiences, and especially intentional experiences, are a consciousness *of* something. Therefore they are temporal, although not spatial and not real (337). By describing the epoché as an abstraction from embodied consciousness Husserl is, according to Smith (whose interpretation I am following here), defining the concept of a pure consciousness. However, it is only possible to articulate this notion in conceptual activity, and never in reality (356). "Pure consciousness is thus an abstraction from embodied consciousness in nature and encultured consciousness in the life-world: an abstractable moment or dependent part of the psychophysical *I* and the human *I*. The process of abstraction is epoché. But the mind is not ontologically separated from the body, as in Cartesian dualism" (351). By bracketing the object of its experience—by bracketing the whole natural world—the ego comes to focus on the way the object is given in the experience (332). By bracketing the question of the existence of the natural world around us, we turn our attention to the structure of our own conscious experience. We thereby recognize that each act of consciousness is intentional, or a consciousness of something. "With the being of the natural world in brackets, all I can say about the tree before me is that in my perspective in this act of seeing that tree, the tree has being *for me* in my seeing it. The tree might or might not exist in itself" (383). We bracket the concrete spatio-temporal event, or one might say, as Smith does, that we abstract from it, we leave it aside and concentrate on the structure of our conscious experience. The goal of that method is to get at consciousness as a domain that has an existence besides nature, but has this existence in an ontological sense not without nature.

Before calling this procedure into question, I will try to clarify its aims, of which there are two. The first is to find the apodictic ground Husserl was seeking, the one that absolutely excludes, as he says, every possible doubt.[11] The second is to sublimate the concept of subjectivity as a ground for the exact sciences and for the life-world.

The exact sciences are, as Husserl says, the accomplishments of the consciousness of knowing subjects. Subjectivity and the accomplishments of consciousness are therefore for Husserl that immanent reason (*die immanente Vernunft*) without which the objectivity of the sciences becomes absurd (*widersinnig*). As they are not domains of a priori truths, they derive their inherent reason from subjectivity. Subjectivity and consciousness are therefore *prior* conditions for the sciences and for the life-world. The immanent reason that subjectivity and consciousness impose on the sciences and the life-world is also part of the ground that enables a critique of the sciences.[12]

We have already seen that the recourse to subjectivity is not sufficient as a ground for critique in the *Crisis*. But maybe the concept of the life-world, which is also determined as a realm of subjective phenomena that have remained "anonymous," can help us in reconstructing historically anonymous subjective *phenomena* as the immanent reason of scientific developments.[13]

> And did this not imply that they all repose upon *one* single ground, one to be investigated scientifically in advance of all the others? And can this ground be . . . any other than precisely that of the anonymous subjectivity we mentioned? But one could and can realize this only when one finally and quite seriously inquires into that which is *taken for granted,* which is presupposed by all thinking, all activity of life with all its ends and accomplishments, . . . This implies first of all the mental accomplishments, which we human beings carry out in the world, as individual, personal, or cultural accomplishments. Before all such accomplishments there has always already been a universal accomplishment. . . . We shall come to understand . . . the world . . . as the unity of mental configuration, as a meaning construct (*Sinngebilde*) as the construct of a universal, ultimately functioning subjectivity. It belongs essentially to this world—constituting accomplishment that subjectivity objectifies itself as human subjectivity, as an element of the world. (*Crisis*, § 29)

The sentence "We shall come to understand . . . the world . . . as the unity of mental configuration, as a meaning construct (*Sinngebilde*)—as the construct of a universal, ultimately functioning subjectivity" could just as well have been written by Hegel. Husserl is trying to reconcile the notions of subjectivity or consciousness, the notion of the contingent world and the notion of reason. He tries to link immanent reason to the realm of subjective *phenomena.* Does this make sense? Does it make sense to try to reconstruct historically anonymous subjective phenomena as the immanent reason of scientific developments?

Consider the following historical example of subjective phenomena, which is also a case for the history of the notion of consciousness. If we study the notion of *thymós* (*heart*, *spirit*, or *anger*) in Homer's writings

(*Homer*, 1925, 23.370),[14] we can see that the strongly pulsating thymós is a bodily agitation, which is connected not only with emotions like anger, but also with thinking, intention, and imagination (Rappe, 1996, 218–19). There is no separation between bodily agitation, phenomenal feeling, and thinking as we know it. The division between immaterial being and its bodily basis presupposes an understanding that splits the soul from the body (Arbman, 1926, 168). This understanding has been prepared by the cult of Dionysus and was continued by Plato (Schlesier, 2000), although even for Plato, consciousness (or the soul) is not something internal that can be separated from the body, rather it is the individual and his or her life. There are phrases like those in Homer's *Odyssey* (19.454) which indicate that life leaves the body when it is fatally injured, but this is not a sign that the notion of consciousness is available to this way of thinking. It just points to the circumstance that the body ceases to be alive when it is fatally injured. If the conventional interpretation of Homer is right, then it is clear that there is no intention, no thinking that can be "bracketed" for the contemporaries of Homer. They could not have had access to an absolute ego, and this indicates that the notion of an absolute ego is related to a historical time. These considerations indicate that the concept of consciousness has a history, and as this was at least already known in the 1920s, Husserl could well have been aware of this fact. Indeed it appears he was aware of this, as the following quotation of the *Crisis* shows very clearly:

> What the modern period calls the theory of the understanding or of reason—in the pregnant sense "critique of reason," transcendental problematics—has the roots of its meaning in the Cartesian *Meditationes*. The ancient world was not acquainted with this sort of thing, since the Cartesian epoché and its ego were unknown. Thus, in truth, there begins with Descartes a completely new manner of philosophizing which seeks its ultimate foundations in the subjective. (*Crisis*, § 19)

So what follows from my assertion that Husserl was very much aware of the historical origin of the concept of consciousness? Husserl has to claim that the epoché is applicable *to* the Homeric age but not *for* Homer's contemporaries. In other words, *we* might apply it to Homer's phenomenal subjectivity, but Homer himself would not have been able to do so. Thus Husserl has to suppose that there is something like a historically evolved concept that stands for an absolute concept. The notion of an absolute ego is applicable, for Husserl, to Homer's individuals. As Smith puts it, "Pure consciousness is . . . an abstraction from embodied consciousness in nature and encultured consciousness in the life-world . . . . The process of abstraction is epoché. . . . This process is only possible in the conceptual activity, and never in reality" (1995, 351, 356). But on my view, the epoché would not have been feasible even as a conceptual activity for Homer. He would not have been able to perform the abstraction or bracketing that is needed, because to do so, one already needs that concept of subjectivity that starts to evolve only with Descartes.

The reduction to a pure intention or absolute ego is thus a process that is dependent on a historical development. Even if we conceded that the epoché was feasible for us, it would still depend on a historically evolved concept. An absolute ego is only conceivable for those who live in a culture that allows for such an abstraction. If we now recall that subjectivity and the accomplishments of consciousness are, for Husserl, inextricably linked to immanent reason, we should conclude that this kind of immanent reason is only available for those living after Descartes. But how could there have been a life-world as a realm of subjective phenomena that remained anonymous before Descartes?

I would suggest that this is only thinkable from a Hegelian perspective. Only if we suppose a historically developing subjectivity that evolves at a particular time in an absolute ego might we understand what immanent reason could mean (without having pure consciousness in a Kantian understanding at hand). The immanent reason must then be one that is relative to historical developments, as well as being relative to the subjective phenomena of a historically remote life-world. One does not have to insinuate that Husserl was an expert in the writings of Hegel in order to read him in this line, but one can take it for granted that every German student in the 1930s knew that much about Hegel. The implication is that Husserl had not only Kant but also Hegel in mind when he wrote in the letter to Albrecht I cited earlier "that all of the great thinkers of the past, whom I admired so much, and whose thoughts have grown in mine in new kinds, had to rank me as a true heir of their spirit, as blood of their blood."

From such a Hegelian perspective, we would have the following view: There is an immanent reason of history that is linked to subjectivity and consciousness. But as the immanent reason is linked to historical development, subjectivity and consciousness can only get to that point after an absolute ego is conceivable. In Husserl's reflections, this particular time is the time after Descartes, because Descartes' successors are able to distinguish between immaterial being and its bodily basis in a conceptual activity. The life-world, as a realm of subjective phenomena existing before Descartes and that is supposedly accessible without its encultured consciousness, is actually accessible only for his successors. The accessibility is guaranteed by, on the one hand, that immanent reason that binds us to the grounds of original evidences, and, on the other, by that pure consciousness in which the immanent reason accomplishes itself. If, by contrast, one does not share such a Hegelian philosophy of history, one has to conclude that the tensions between ahistoricity or nonhistoricity are not resolved in the *Crisis*.

§ 9 Science, Philosophy, and the History of Knowledge: Husserl's Conception of a Life-World and Sellars's Manifest and Scientific Images

Michael Hampe

Both Husserl and Sellars have ambitious programs for philosophy; they each consider philosophy as the main cultural force. For them philosophy has a historical mission, and this mission is not defined by a moral or political goal, but by a theoretical or epistemic task. Philosophy has to create the unity and ultimate transparency of human knowledge. The epistemic ideals of unity and transparency are old and were first articulated by Descartes. In this chapter, I will discuss how both Husserl and Sellars want to realize these ideals using a certain type of history of knowledge. Their conceptions are comparable because they both use similar conceptual dichotomies to characterize the difference between scientific and nonscientific knowledge. In the second part of this chapter, I will criticize these attempts as not doing justice to the complexity of the social and emotional setting in which knowledge develops.

Introduction

The philosophical accomplishments of Husserl and Sellars are great. Phenomenology, Husserl's philosophical creation, has become one of the strongest intellectual traditions of the twentieth century in philosophy, sociology, and cultural anthropology, and it was the starting point of philosophical schools like existentialism. The power of Sellars's transformation of the philosophy of the late Wittgenstein has only just now become entirely obvious in the inferential semantics of Robert Brandom. In the following, I will not deal with these obvious credits to Husserl's and

Sellars's thought, but with something that seems to lie more on the periphery of their work.

It is the merit of both to have seen that for a "complete" view of the world, fact and values, objectivity and subjectivity need to be brought together. This is an insight that is present today especially in the neo-pragmatism of Putnam. But Husserl and Sellars also recognized that there is a tension and a conflict between our search for facts and objectivity and our desire to understand values and subjective perspectives. They gave names to the sides that produce this tension: *the manifest image* and *the scientific image* (Sellars) and *Galilean science* and *the life-world* (Husserl). In their awareness of the conflicts in modern Western culture, they were perhaps more advanced than some neo-pragmatists. In this chapter their view of the difficulty of relating scientific and nonscientific knowledge is not only analyzed but also criticized for being still too simple. This criticism is not meant to be destructive. Although their general diagnosis is right, it will need much more sophistication and is to be connected with a humbler view of the cultural possibilities of philosophical texts in order to lead to some "therapy."

Galilean Science, the Life-World, the Manifest, and Scientific Images

Husserl and Sellars see a disunity and opacity in the knowledge of their own times: a disunity between a science dealing with facts and the "problems of reason," as Husserl calls them, in which values and meaning are to be found (*Crisis*, § 3). Connecting facts with ideals, values, and meaning (*Sinn*) is the primordial task of philosophy, according to Husserl. In a very similar spirit Wilfrid Sellars writes in his "Introduction to the Philosophy of Science" from 1964:

> There is some measure of truth in . . . [the idea of a unified science], but it overlooks, among other things, another dimension of philosophy. We want to understand in philosophy not only what is the case and how the world of fact operates, but we also want to understand how this relates to human living, human values and obligations, to the experience of beauty and to religious experience. Thus, even if the sciences are integrating themselves, there still remains the task of seeing how the world of facts and the world of value fit together. (1964, 3)[1]

Sellars does not question the idea of a unified science, but thinks it will not solve the problems of disunity that the advocates of a unified science believe it would solve. The disunity between fact and value, or fact and meaning, results from the supposed incompatibility of two realms of knowledge: the realm of scientific knowledge and the realm of nonscientific knowledge. Husserl's life-world (*Lebenswelt*) as the realm of sense or meaning serves a similar function in his philosophical setting as does what Sellars calls the "original image" and the "manifest image" in his essay "Philosophy and the Scientific Image of Man" (1962). The unification of the realm of ideals with the realm of fact under the intellectual leadership of philosophy will bring European mankind, according to Husserl, to its final goal. Therefore philosophers are for Husserl "functionaries of mankind" (*Funktionäre der Menschheit*), who carry the responsibility for the "true being of mankind" (*wahre Sein der Menschheit*), which is to be found in the telos of a final state of man, where the dichotomy between fact and value has been dissolved and knowledge has become entirely self-transparent (*Crisis*, § 7). This telos is an ideal for the never-ending phenomenological work required under Husserl's conception of historical reconstructions of the meaning of scientific terms, all of which for him are initially derived from what one would usually call ordinary experience. Following Husserl, in this chapter I will refer to these ordinary experiences as the life-world.

For Sellars, a possible final state of the development of knowledge, in the sense of a fusion or synoptic vision of the scientific and the manifest images, seems to be a real possibility. But today, statements about a final goal of knowledge reached via philosophy appear to be an enormous overestimation of the cultural role and force of philosophy, and these statements are as hard to understand as their teleological background. But as is well known, Martin Heidegger, and especially the late Heidegger, not only took over but intensified this overestimation of the cultural mission of philosophy. This is evident in his texts about the relevance of thinking and in those where he suggests that the whole history of mankind, especially the history of science and of technology, as far as it is influenced by Europe, are the result of *Seinsvergessenheit*, the forgetting of Being; that is, for him history is interpreted mainly as a history of philosophy with different epiphenomena. For example, technology appears in this particular history as a form of metaphysics (*Gestalt der Metaphysik*).

Sellars does not talk about mankind. But he says that man created himself. When man produced the manifest image, he made it possible to distinguish between things and persons. According to Sellars, this distinction disappears in the scientific image. So the question of whether the scientific image and the manifest image could be brought into a fusion is for Sellars the same as the problem of whether man will stay in charge of his self-productive powers, of his ability to reproduce himself as a reflective and responsible creature by producing a view of himself and the world. In considering the origin of the distinction between categories applying exclusively to man himself as a rational being and categories applying to the world only, Sellars seems to come close to the Fichtean idea of a self-creation (*Selbstsetzung*) of man. At the time when man—or perhaps one should say speechless *Homo sapiens*—was just behaving without having a categorical framework to describe and explain his behavior, he did not really exist as man. By producing the category of *man* and the terminology about inner states, man brought himself with "a jump" into being, according to Sellars. He writes:

> The manifest image of man-in-the world . . . is . . . the framework in terms of which man came to be aware of himself as man-in-the-world. It is the framework in terms of which . . . man first encountered himself—which is, of course, when he came to be man. I have given this quasi-historical dimension of our construct pride of place, because I want to highlight from the beginning what might be called the paradox of man's encounter with himself, the paradox consisting in the fact that man couldn't be man until he encountered himself. . . . The conclusion is difficult to avoid that the transition from pre-conceptual patterns of behaviour to conceptual thinking was a . . . jump to a level of awareness which is irreducibly new, a jump which was the coming into being of man. (1963, 6)

Fichte was not talking of a jump but of a fundamental action (*Tathandlung*). But his idea seems very similar to what Sellars says. The problem we are confronted with here is one to be found throughout the philosophy of German idealism from Fichte to Hegel. All philosophers of this "school" try to give an account of how some very fundamental human capacity is possible. In giving their account they use a "genetic" terminology, which gives a nondeterministic picture of *x* evolving or becoming or changing and thereby bringing to light *y* and *z*. For example, in Fichte's *Wissenschaftslehre*, human beings bring about free man by a primordial

Tathandlung, and nature gives rise to consciousness by an "evolution." In Schelling's *System des transzendentalen Idealismus*, consciousness develops into self-consciousness. And understanding develops into reason und absolute knowledge in Hegel's *Phänomenologie des Geistes*. Each says these genetic accounts are not meant to be descriptions of processes in time, but transcendental deductions or dialectical developments of concepts. But deduction does not necessarily mean logical inference as we commonly understand it. So the reader is left with some kind of formal, but not logically formal, genesis, a philosophical history of some important human capacities that may not relate to the real history of man. In Hegel this process can perhaps be interpreted as an idealized educational development, comparable to what Bachelard might have imagined when he wrote his theory of the development of the scientific mentality. But it is very hard to understand what stands behind these types of genesis and whether they are to be distinguished from mere imaginations, because references to historical sources are not relevant here. In Kant the term *transcendental deduction* can be interpreted in a legal sense. In Fichte, I believe, this is not possible. It seems to me that both Husserl and Sellars try something similar to these formalized types of genesis and therefore are of comparable obscurity in their philosophical histories.

Synchronic and Diachronic Priority

The realms of scientific and nonscientific knowledge are related to each other in two ways: by a *diachronic* link and a *synchronic* one. In a diachronic link, the life-world stands historically at the beginning of all scientific developments, but this link can conflict synchronically with scientific opinions. For instance, if one looks at an object and believes that it does have color in the life-world, but physics tells us that these objects do not in fact have color, then we have a synchronic conflict between our experience and the supposedly resulting scientific theory.

In Husserl, the realm of Galilean science is historically later than life-world experience, and in Sellars the original image and the manifest image are historically earlier than the scientific image. For both, the life-world is systematically and historically the source of all meanings that the naturalistic sciences are unable to produce for themselves. Therefore for Husserl the life-world has a systematic and historical *priority* over any scientific theory or worldview. For Sellars the manifest image develops out of what he calls the original image by a "categorical differentiation," and the scientific image develops out of the manifest image. Sellars puts the priority of the manifest image above the scientific image, and this priority is both systematic and historical: The scientific image is to be reconstructed against the background of the manifest image, and it is a historical product of a particular development of the manifest image. The original image was a myth. In it no distinction is made between intentions and actions on the one hand, and causes and effects on the other. It is, in the strict sense, not an image of man in the world, because man—as we saw in the passage quoted earlier—does not exist for himself in this image. Here man has no category for his intentional and reflective being, especially not for his rule-following capacity to explain something in a "conceptual scheme."

The manifest image alters this by introducing theoretical and practical terms, or terms for "things" on the one hand, and for "personal categories" on the other: Everything that is to be explained by the inner states of a person belongs to a different categorical framework than what happens independently from personal states. According to Sellars, the invention of the manifest image is the invention of man as an independent reflective and practical being, who can follow rules of reasoning and produce justifications for actions. The scientific image, as the latest developmental state in Sellars's genealogy of images of man in the world, wipes out this distinction between personal states and natural events and tries to understand human affairs as special cases of the concatenations of natural causes and effects. So in a sense, the scientific image leads back to the original image, in which the difference between man and the world did not exist as a categorical one. The homogeneity of the scientific image seems to be for Sellars the preliminary prize for the enormous explanatory success of the natural sciences, a success not to be found in the original image, which was the reason for its historical instability. The instability of the constellation between the manifest image and scientific image that Sellars is diagnosing comes not from explanatory failures but from the *incompatibility* of these two conceptual frameworks.

Synchronically, the two realms are connected with each other by man, or in Husserl's terms, through "the European form of man" (*das europäische Menschentum*). Human beings as scientists do not change their worlds or their consciousness when they leave the breakfast table and walk into their laboratories. They have to bring together, somehow, what they believe as scientists and what they believe as wives and husbands at the breakfast table. It is both the supposed unity of consciousness and the supposed fundamental self-givenness of a transcendental ego in Husserl, and the supposed unity of our language in Sellars, that force us to search for a connection between facts and values, the scientific worldview and the life-world or the manifest image. The idea that human beings might think in compartments or play different language-games and might not care very much how the compartments and language-games hang together are not options for Sellars or Husserl, but relativistic nightmares. To opt for such a relativism would be considered in their individual frameworks as opting out of European rationality. Both connections—I will call them the historical and the anthropological one—have their complications.

Problems from the History of Knowledge

Before I discuss these complications, I should point out that it is very difficult to understand in what sense these realms, looked at separately, are unified entities and what the possible criteria for their identity could be. As can be seen in the first quote from Sellars, he thinks that the sciences aim at unity, but that they do not yet have this unity. Unity of the sciences is considered by Sellars to be a historical goal. If one considers the sciences as systems of theories, the unity of scientific knowledge would be, in a weak sense, the compatibility of scientific theories. In a strong sense, the unity of the sciences would consist in the reformulation of all scientific knowledge into one language, their integration into one grand theory.

However, the sciences consist not just of theories but also of technologies and practices, which transport know-how, in other words nonpropositional knowledge. In the actual historical process, one does not only see a multiplication of scientific terminologies but also of technologies and scientific practices. For Husserl the technologies and practices of the sciences are opaque knowledge. Even the use of mathematical shortcuts in physics or engineering is for him unscientific, in that they impart no knowledge in the sense of Aristotelian *episteme*.

The phenomenological search for transparency is to be interpreted, it seems to me, as the infinite project of making explicit everything in our knowledge, so that we have at least a theoretical possibility of understanding any technology and conceptual or mathematical shortcut by looking at the historical reconstruction of its origin. But the growing body of tacit knowledge in the sciences arising from the differentiation of practices and technologies makes it difficult to see in what sense there can be a hidden tendency for unity behind the development of the sciences as Husserl and Sellars seem to believe. Furthermore, if one distinguishes between propositional and non-propositional knowledge, one has to observe that the nonscientific realms of knowledge contain much more non-propositional, or implicit and tacit, knowledge than explicit knowledge. What would the unity of a realm of practices in a life-world or in the manifest image be? It cannot be the unity of a coherent theory, that is, the unity of a set of propositions, because the tacit knowledge and the practices of the life-world, or the manifest image, are never given to us in a theory. If there is an expression or description of this tacit knowledge at all, it can, for example, be found in social history about certain epochs and in novels.

I see no theoretical criteria for the unity of these non-propositional knowledge-structures in Husserl or Sellars. Using this background, is it possible to imagine a unity of the knowledge that is implicit in practices like rowing, dancing, writing, and singing with the practices followed in laboratories like reading an X-ray picture or a smear or cutting a piece of mouse brain with a microtome? Husserl's "form of life," or the "comprehensive style" (*Gesamtstil*) of seeing the world, and Sellars's "image" both seem to suggest that the beliefs and practices of the nonscientific realm come together into some kind of unification that is independent of the inferential structures of theories (*Crisis*, § 9b). What would be the principles of unity here, and how can we individuate a form of life, a style in which to see the world or a manifest image? I see no answers to these questions in Husserl and Sellars, and it is probably no accident that the terms *style* and *image* are rather vague names for the unities they are considering here.

Sellars says that the manifest image and the scientific image are "ideal types" for him in the sense of Max Weber (Sellars, 1963, 5). That statement throws some shadows on the idea that there was a historical development from the original image, then to the manifest, and finally to the scientific image. Can there be a historical development from one ideal type to another? Ideal types are comparable to epochal terms like *Renaissance* or *modernity*; they are systematizing terms. These terms are constructed in order to produce a simplifying structure in the hopelessly complex situations that are presented to historians by their historical sources. But considering things in this historical manner almost always relativizes such systematizations by revealing supposed continuities and particularities that do not fit into the system. Odo Marquard once pointed out this fact by saying that we live in the epoch of a deflation of epochal thresholds (*Epochenschwellenabschwellungsepoche*).

Husserl believes that the opacity of the sedimentation of knowledge can be made transparent again by a special kind of phenomenological historical research; research that will not use conceptual techniques but will reconstruct their origin. But this reconstructive work cannot be done without a historical technique. The historian of knowledge needs as many conceptual tools as any other researcher who has to organize his material. As soon as these conceptual tools of historical research are taken to be a reality, the researcher falls into the trap of the fallacy of misplaced concreteness (as Whitehead had called it). It is a nonstatement, or a manifestation of this fallacy, to say that the Middle Ages developed into the Renaissance and the Renaissance into modernity. This is playing with one's own systematizations, but it is not producing historical insights or explanations. Husserl and Sellars do not produce such nonstatements. But the tension between a historical approach to knowledge and a systematizing approach is as characteristic of the Husserl of the *Crisis* as it is for Sellars (1962) in "Philosophy and the Scientific Image of Man."[2]

By considering that the historical task of philosophy is to unify the realms of the life-world and the manifest image on the one hand and the scientific worldview on the other hand, both philosophers take a peculiar methodological standpoint in the historiography of knowledge. The history of knowledge is guided for them by a goal and a source. For them, as for Bachelard, the role of the historian is not just the role of an observer. In their view, the historian of knowledge *participates* in the development of knowledge, not by producing a special kind of knowledge, but by knowing where all knowledge comes from and where it aims to go. The "teleological motor" in the history of knowledge may be different in different participating historical projects. It is the constantly differentiating plurality of the sciences that needs a plurality of epistemological investigations: those the historian must work out for himself in Bachelard and the required unity of reason and the sciences in Husserl and Sellars.

According to these views, the historian of knowledge has to push knowledge in the right direction by using his historical systematizations. Because of their understanding of the history of knowledge, Husserl and Sellars each have a clear vision of the origin of these worldviews and images, and they seem to possess strong convictions not only about the future unified state of the sciences but also about the future of knowledge as such.

Husserl believes that the primal establishment (*Urstiftung*) of the European form of man (*europäisches Menschentum*) is connected with the ideals of freedom and of science (*Crisis*, § 5). Science as a form of life could only develop where man becomes conscious of himself as a free being, who can produce his own way of living and is not forced to live in a certain way. Asking for reasons for everything and not believing in authorities is an essential feature of what Husserl believes to be the European style of living and of science. Because science and the life-world, with its personalistic view of free self-responsible man, grow from the same historical root, they must, Husserl seems to argue, be compatible and unifiable in some way and at some time. A fundamental origin and a final goal of the historical development go together for him: "But to every primal establishment (*Urstiftung*) essentially belongs a final establishment (*Endstiftung*) assigned as a task to the historical process" (*Crisis*, § 15). Sellars believes that both the manifest image and the scientific image have the same function: They try to explain what happens in the world and why persons behave the way they do behave. Because both images are "produced" for the same explanatory aims, it must be possible, so Sellars seems to argue, to fuse them into what he calls a "synoptic vision." Thus an ultimate origin and a common function serve as promises that a unification of life-world or the manifest image and the scientific worldview is possible. This idea of a unification of scientific and nonscientific knowledge leads to very special philosophical projects.

Husserl tries to make the life-world transparent in a way that gives a basis for all human activities, and is constitutive for science, through his project of transcendental phenomenology. This phenomenology is, according to Husserl, the hidden aim at which the history of the European form of man is directed from the start. Sellars, however, wants to anticipate a process of metaphysics as the hidden aim of the sciences. A metaphysics that will dissolve the contrast between a person, as an entity with intentions and a life-history on the one hand, and a physical thing, as an opaque entity with more or less no history on the other. Processes have, or perhaps even are, histories, and they may have protentional and retentional phases that would make it possible to integrate human reflectivity with natural causality.

According to Sellars, it is this contrast between persons and things that at present keeps the scientific worldview and the manifest image apart. In a metaphysics of pure processes, as Sellars sketches it in his Carus Lectures (1981), histories of mental events will no longer be reconstructed as exotic entities that can only be explained in an ontology that takes the concept of a person as fundamental. Pure processes and their temporal structures are the elementary entities in the reconstruction of states of consciousness and also, Sellars thinks, the starting point of the reconstruction of any scientific object. The justification for such a speculation can be found, according to Sellars, mainly in the tendency of modern physics to take processes and fields to be more fundamental than enduring objects like atoms or substances. (Here Sellars seems to take the quantum field theory somehow as the most fundamental reductive basis of all physical theories.) The unification of knowledge will happen, for Sellars, from the standpoint of an ideal science, which revises our view of enduring individuals like persons and things in one move.

Husserl, on the other hand, takes a different approach to a unifying conception of reality or a universal knowledge. It is not the anticipation of a future fundamental scientific ontology but a philosophical analysis of the life-world and a historical phenomenology of the sciences that applies in his analysis. While Sellars produces a kind of heuristics for a possible integrated scientific view in the future, one that would not stand in opposition to the self-understanding of man as it is given now in the manifest image, Husserl believes that the analysis of the life-world will again bring to light the fundamental meanings that made science possible. Thus for him it is not heuristic metaphysics but the history of science that is the key project that will help us attain the transparency and unity of knowledge.

Four Critical Remarks

I would like to conclude with four critical remarks on these projects of Sellars and Husserl:

1. The realization of the epistemic ideal of the transparency of knowledge takes tacit knowledge as an epistemic problem that is to be resolved through historical inquiry. Why should we do that? Why should we think that somebody, say a craftsman or a scientific experimenter, is in an epistemically imperfect state, in that he knows what he does but can give no description of it? Certainly, the idea that explicit contemplative knowledge, laid down in minute descriptions, is a superior kind of knowledge has existed since antiquity. But do we really still share this epistemic ideal in the way Husserl seemed to have felt committed to it? And why should we share this view? If we do not share it, then we will lose one reason to do history of knowledge in the spirit of Husserl's *Crisis*. In one sense the life-world seems already to be a basis, or "fundament," of all scientific knowledge for Husserl. But as long as a phenomenological history of the sciences has not brought the meaning of scientific terms back to their life-world roots, we do not see this fundament. A fundament of knowledge that is not known as a fundament is not an epistemic one for Husserl. He seems to have an entirely different view of the tacit, or practical, dimension of human knowledge than does, for example, the Wittgenstein of *On Certainty*. Husserl wants to reconstruct the unknown, untransparent fundament of all knowledge in a reconstructed life-world. But what would we really gain in rationality if the project of making everything explicit were possible and were subsequently undertaken? Why should we consider the infinite task of striving for ultimate transparency as more rational than being content with the general insight our knowledge always has provided on a practical basis, despite our not knowing this "fundament" in its entirety?

2. Self-transparency of knowledge through a history of knowledge would only be possible if the history of knowledge did not itself work with implicit knowledge and used only those categories that are understood in their historical context. I do not believe that any historian of science can evaluate sources like letters, laboratory notebooks, article drafts, instruments, and so forth without any tacit knowledge about these cultural artifacts, a tacit knowledge that the historian will gain by working with these kinds of sources over a long period. Furthermore it is unclear to me whether Husserl's terms life-world and style, along with his category of a scientific or naturalistic view of the

world, should not themselves be described as conceptual shortcuts, as tools to systematize historical material, which would only engender chaos if not structured by such means. Husserl's idea of the transparency of knowledge through a history of knowledge seems to be closely connected with the conviction that there can be a phenomenology of historical facts, that is, that there can be an *immediate* knowledge about historical data. I do not believe that there is such a thing.

3. Why does Sellars use the term image for his knowledge structures and not the Wittgensteinian term *language-game*? Sellars is in fact in many of his arguments a Wittgensteinian. But if he had been Wittgensteinian in the description of what he calls the manifest image and the scientific image, he would have lost his dualistic contrast as an engine for the history of knowledge. We play many different language-games outside the realm of the sciences. And we play many different language-games in the sciences. It is very easy to smooth the connection between explicit and tacit knowledge in the concept of a language-game. In a sense, this Wittgensteinian term was made to grasp this connection between explicit and tacit knowledge or knowing how and knowing that. But then the relations between different language-games are not very useful for historical projections like the ones we have seen in Husserl and Sellars. Certainly, language-games also come from somewhere, and they are developing into something. But probably nobody would dare to think of a primordial source of language-games and a final goal where they all come together into one big game. Such a relativistic tendency, which one can also find in Cassirer's philosophy of symbolic forms, would lead to an entirely different view of the history of knowledge, a view that would consider theories as one, but only one, manifestation of knowledge; crafts, novels, and technologies would also be other kinds of nontheoretical knowledge.

4. What we think about ourselves and our fellow humans is not so much a matter of theories and propositions but of attitudes and emotional habits. We have certain feelings of shame and regret, pride and pity, about ourselves and other people, not because we live in a life-world or follow a scientific view of the world, but because we act in certain institutions like families, schools, law courts, and so on.

Even if the scientific image or the naturalistic worldview ceased to be challenged by argument, we would probably not alter these institutions. Certainly, science has an influence on families, schools, and law courts. But I dare say that the way we see and feel about human beings is more constantly determined by these institutions than by what the sciences say about man. What Husserl and Sellars have to say about the relevance of scientific knowledge and the life-world or the manifest image is much too simplistic if one takes these social institutions into account. It does not do justice to the complex way in which a culture forms a way of understanding and reacting to human beings through its institutions. It seems to me that the ideas of a life-world and a manifest image are even dangerous, in the sense that they give us the illusion that in philosophy we know how human beings come to understand themselves, without looking into the institutional, emotional, and cultural details of certain societies. In this sense, both notions could foster a kind of anthropological essentialism and fundamentalism that would be of little use for a historical epistemology that is interested in historical details.

§ 10 On the Historicity of Scientific Knowledge: Ludwik Fleck, Gaston Bachelard, Edmund Husserl

Hans-Jörg Rheinberger

In this chapter, I look into an intellectual and political debate that was fought during the first decades of the twentieth century under the banner of a "crisis of reality"—a battle equally perceived as a "crisis of historicism." This double crisis also had consequences for conceiving the history of science, which experienced a fundamental change in the period between 1880 and 1930. The history of science abandoned its roles as the chronicler, the eulogist, and moralist of the sciences and emerged as a field in which the primacy of positivism (for the sciences) and historicism (for the humanities) began to be questioned. In what follows, I will argue that the development of a history of science with a genuinely historical-epistemological agenda is intimately connected with the crisis experienced at the beginning of the twentieth century.

In 1872, the distinguished Berlin physiologist Emil Du Bois-Reymond lectured *On the History of Science* at the Leibniz Session of the Academy of the Sciences in Berlin. In this lecture, the idea of a history of science as a means of teaching and understanding the present state of a discipline is preeminent. Consider the following quote from Du Bois-Reymond's paper: "No matter whether we deal with an organism, a state, a language, or a scientific doctrine, it is always its developmental history that best captures the meaning of and the relation between things." And Du Bois continues: "In contrast to the dogmatic representation that is often used in textbooks, I prefer an inductive approach. In the textbook as well as in the lecture hall, this is the right way to teach physiology. In this manner, one teaches the science and its history at the same time" (1912, 432–33, 436). This is a good example of the rather unproblematic and quasi-natural uses of history of science before the critical turn. For Du Bois-Reymond, it appears evident that besides "the contingencies of the business of discovery," "the historical path of the inductive sciences is mostly the same as the path of induction itself" (1912, 435). Indeed, here we may even talk of a "natural history" of science, a history following the unfolding of an inductive logic. At the same time, this attitude is part of a pervasive historical perspective not only in the humanities but particularly also in the life sciences of the late nineteenth century. There is a peculiar reciprocity to be observed here: An extensive historicization of nature runs side by side with a naturalization of history.

Physicists such as Ludwig Boltzmann did not hesitate to proclaim the nineteenth century, not as might be expected, as the century of steam power or of electricity, but rather as the "century of the mechanical conception of nature, the century of Darwin" (Boltzmann, 1905, 28). This connection between mechanics and natural history reveals that scientists such as Boltzmann did not take Darwin's theory of evolution at all as an indication of a genuine and extensive historicization of nature, including life. On the contrary, Boltzmann took Darwin's achievement as living proof and evidence of the possibility of a *mechanization* of history, a mechanical—and may we add, statistical—understanding of nature, including the genesis of living beings.

With this couple of quotations, I wish to draw attention to the fact that at the turn of the twentieth century, an extensive historicization of nature—including life—fits in very well with an overall mechanistic explanatory drive, including a naturalization of history, even resulting in a "decreased interest of the scientists concerning the historicity of their own knowledge," as Gregor Schiemann has observed (Schiemann, 1997, 157).

Change came, however, from within the development of the sciences themselves. Two phenomena can be discerned that, during the first decades of the twentieth century, begin to resist the drive of mechanical thinking. They are not alone responsible for, but contribute decisively to, an intellectual mood and new meaning of historicity that I will describe in more detail in the second part of the chapter, in which I discuss the works of the young physician Ludwik Fleck and those of the aged philosopher Edmund Husserl—with a few interspersed remarks on Gaston Bachelard. These two phenomena are, first, the revolutionary developments in physics, and second, the problem of the unity of the sciences.

The first is well and widely known. I will not enter into a discussion of relativity theory or quantum physics. It shall suffice, as a brief remark, to note that two insights resulted from the revolutionary changes in physical theory. Both of them had consequences for thought and self-perception in the history of science. I will report on both of them—pars pro toto—as articulated by the contemporary protozoologist Max Hartmann. As to the first point, Hartmann contends that one must "keep one's eyes and one's brain open for the discovery of new conceptual systems, if experience demands them." One has to take to heart the admonition "that it is dangerous for thinking and for science, if one generalizes the results of an epoch such as those of the high noon of classical physics with too great a dogmatic security to other domains of the sciences and the humanities and even to the whole intellectual world. For that means transcending the competences and the boundaries of one's own scientific discipline" (Hartmann, 1956, 155).

In these remarks, a new consciousness regarding the historicity of scientific knowledge appears. Classical mechanics had not become obsolete through the new developments. First, however, it became realigned with respect to its own boundedness in time; second, it became located as a historical step in the epochal development of physical knowledge; and, third, it became recognized in its own phenomenal restriction. Here scientific knowledge is no longer experienced as being directed toward closure and perfection, but rather as an "infinite progress" in the words of Hartmann, a progress whose horizon is no longer given once and forever, any more than the direction it will eventually take. Hartmann resumes rather categorically: "Every statement about the nature of world reality—*Beschaffenheit der Weltwirklichkeit*—is a historically conditioned one" (1956, 150). Similarly and around the same time, in his *Essai sur la connaissance approchée* of 1928, Gaston Bachelard in France states the point in the following words: "In our view, this reality, in its inexhaustible unknown, exhibits a character that lends itself eminently to a research without end. Its very essence resides in its resistance to being known. We take it thus as a postulate of epistemology that knowledge is unfinished in a fundamental fashion" (Bachelard, 1987, 13). Furthermore, Bachelard also stresses the moment of the unprecedented and the unanticipated in the course of knowledge acquisition: "The history of the sciences teaches us that every great progress toward an ultimate reality has shown that this reality found itself in a direction that had by no means been expected" (284).

With that, I come to the second of the above-mentioned points, namely the problem of the unity of the sciences. Even biologists such as Hartmann, who definitely stuck to the causality principle in his biological work, unambiguously postulated a qualitative difference at least between the realm of physics and that of the biological sciences. He did so within the confines of the natural sciences themselves and without having recourse to neovitalist resources. Hartmann stated: "The task of causal research in biology is not to reduce biological events and processes (*Geschehen*) to physical-chemical reactions. The task is rather the elucidation of *the specific laws of complication* which determine the essence of these particular, individualized natural bodies" (1956, 152). He thus claimed a terrain of scientific research (other than physics) on which other things were sought for and explained. In addition to this principal claim came the division of this other terrain—biology—itself: namely, the factual splitting up of the life sciences into different disciplines that had very different knowledge horizons—from physiology through developmental mechanics to phylogenetics and evolution. That split conveyed the feeling that it was not only biology that displayed an irreducible plurality but the large remainder of the sciences as well. In the very moment when a philosophy of science movement such as the Vienna Circle once again set itself the task of spelling out the unity of the sciences, the genuine developmental dynamics of these sciences themselves appeared to doom these unification efforts to failure.

The demarcation of biology from physics and the internal differentiation of the life sciences into apparently irreducible disciplines have occupied philosophically minded biologists since the beginning of the twentieth century. The relation between physics and biology was usually discussed under the title of "theoretical biology," whereas the internal differentiation of the life sciences resulted in multiple attempts at producing more or less extensive "general biologies," such as Hartmann's *Allgemeine Biologie* (1927). As we have seen in the example of Hartmann, this movement cannot be reduced to a resurgence of vitalism. Of course, issues of wholeness and synthesis featured prominently in these attempts to deal with theoretical as well as general questions of the life sciences during that time; yet, with few exceptions, they remained thoroughly on the grounds of specific domains of experience and corresponding regimes of experimentation, as has been neatly illustrated by Anne Harrington in her *Reenchanted Science* (1996).

In a paper for the journal jointly issued by the Kaiser Wilhelm Society and the Society of German Scientists and Physicians, *Die Naturwissenschaften*, the philosopher and cultural politician Kurt Riezler from Frankfurt summarized the situation in 1928 as follows: "At first, a part of natural lawfulness [*Gesetzmäßigkeit*] that was known to us became unmasked and revealed itself as a mere statistical regularity . . . To this turn, the diverging development of the particular sciences added a second. . . . The individual disciplines did not converge but they diverged; they developed their conceptual systems in different directions" (706).

One year after Riezler's paper titled *The Crisis of "Reality,"* the same journal published a follow-up. It was written by a young, unknown microbiologist and immunologist from Lemberg with ten years' laboratory experience but not yet any credit in philosophy of science. His name was Ludwik Fleck. The article had a title almost identical to that of Riezler. But instead of a definite *the*, we read *A Propos a Crisis of "Reality."* Riezler had distinguished three different "realities": First, the reality of the continuous stream of our exterior and interior perceptions; second, the reality of our objectivizing knowledge of the world; and third, the absolute reality underlying our historically changing knowledge of the world. For Riezler, the development of the sciences of the past decades had eclipsed the belief that our second realities continually approximated the third, absolute one and finally dissolved themselves in it. With that, he stated the crisis he had diagnosed as being an epistemological one, that is, a crisis involving the relation between the second and the third reality. In his essay, Fleck mentioned Riezler's paper only in passing, but with a significant shift in perspective: "If one wishes to solve the problem of the origin of knowledge in a traditional manner as an individual matter of a symbolic 'man,' then . . . there will be no progress and advance. Therefore I do not know why and to what end I should distinguish between a first and a second reality, as described among others by Riezler" (1929, 426).

This remark indicates a displacement entailing fundamental consequences. Fleck, in a first step, reformulates the philosophical crisis in the relation between the second and third realities as a social and psychological one by dissolving the distinction between the first and second ones. Following the above sentence, he explains: "Indeed one must not neglect the social moment in the origination of knowledge." And he adds: "Every thinking individual, as a member of some society, has its own reality, in which and according to which it lives. Indeed, everybody disposes of many, partially even conflicting realities: the reality of everyday life, a professional one, a religious one, a political and a little scientific reality" (Fleck, 1929, 426).

Yet Fleck does not leave it at what might at first glance appear as simple knowledge relativism. In a second step, the socially and psychologically reformulated reality problem is now transformed into a historical one: "Every epistemology must be brought into relation with social matters," he says, "and then, further, with the matters of cultural history, if it does not want to come into severe conflict with the history of knowledge and everyday experience" (Fleck, 1929, 425). Knowledge is therefore no longer conceptualized as the relation of a knowing *ego*, of a "symbolic 'man'" with his object; neither is it simply reformulated as a multiply structured social relation to the world around us, in which the individual is concurrently a member of various, very different social groups and thus moves in different "worlds." Fleck relocates the knowledge problem between the coordinates of everyday experience on the one hand and a cultural history of knowledge production on the other: "For knowing is neither mere passive contemplation, nor acquisition of the only one possible insight into something ready-made and once forever given. It is an active, living engagement in relations, a transformation and a being transformed, in short, a creation [*ein Schaffen*]" (426).

With explicit reference to Niels Bohr's quantum postulate and the non-negligible interference between atomic phenomena and their measuring devices, Fleck states: "Observation, knowing [*Erkennen*] is always . . . literally a transformation of the object of knowledge" (1929, 428). At this point, let us once again briefly switch to Fleck's contemporary, Bachelard. The vicinity to his French colleague, who at the same time, at the end of the 1920s, finds his own way of detaching himself from the positivist epistemological tradition, is striking. Bachelard also recurs to Bohr. However, he steps even further toward an irreducibly historical epistemology. The following is a quote from his *La connaissance approchée*: "Since the phenomenon is absolutely inseparable from the conditions of its detection, it has to be characterized by its detection" (1987, 297). And he elucidates by observing that the source of the scientific river is to be seen as a geographical point only, one that does not contain all its energy: "We are [thus] justified to take knowledge in its plain course, far apart from its sensible origin" (15). Like Fleck, he aims at overcoming what he judges to be "a kind of negative procedure by which one opposes the non-I to the subject." This attitude, at the same time, amounts to a new kind of realism for Bachelard, "a realism without substance," (298) a realism of knowledge as an infinite process and, we can add, a definite transition from a philosophy of knowledge to an epistemology: away from contemplating the relation of the knowing subject to the world it contemplates toward a conception of knowledge as a process that is always, by its means, technically and culturally mediated and invested.

Fleck repeatedly characterized this process in his paper of 1929 as a "democratic" one, that is, as a communitarian enterprise. He saw it shaped by the community of "experts" (*Fachleute*), and the experts in turn were shaped by this same process: "For natural science is the art to shape a democratic reality and to orient oneself accordingly—therefore to become transformed by it. It is a permanent, much more synthetic than analytic, never ending work, like the work of a river which forms its own bed. This is true, living natural science. We never must forget its creative-synthetic and social-historical aspects" (1929, 429). Fleck uses the image of a river here in a manner akin to the phrase of Bachelard mentioned above.

Fleck insisted that the natural sciences should be stripped of what he called their "paper form" and their philosophical transcendence. He contended that "one falsely equates the natural sciences as they are with the natural sciences . . . as one would like to have them" (1929, 427). Once their idealized and ideologized philosophical clothes have been taken away, once they have been brought back to the ground of a real phenomenon (a *Realphänomen*) that can be investigated empirically, they can become the object of historical research, a richly structured cultural phenomenon that by its very constitution embodies a democratic attitude. For it is collective work, and it lives and flourishes from being predicated on preliminarity in the double sense of this word: ready to anticipate, and at the same time, ready to deviate from cherished meanings and to surpass them. It is in this spirit that Fleck assumes the task of showing that the sciences—which were so heavily criticized, especially in the aftermath of World War I, as being mechanistic, devoid of life, if not deadly and disenchanted—that these sciences in their practical, and so to speak, everyday existence, neither function mechanistically, nor are they disenchanted. This is yet another context of the crisis of the mechanical worldview in which the development of Fleck's ideas are embedded. We have to see them as an effort not to discard but to save the sciences from their own cultural image of rigidity. I will come back to this point in the section on Husserl.

Neither Fleck's paper of 1929 nor his later book on the *Genesis and Development of a Scientific Fact*, which was published in Switzerland in 1935, was an immediate success. Fleck's suggestions only became historically effective about forty years after their first publication, namely in the context of the discussion on Thomas Kuhn's concept of paradigm and his view on the structure of scientific revolutions. There are obvious reasons for this belatedness. Fleck's writings were given no time for resonance. His book was ignored in national socialist Germany, and Fleck himself, as a Jewish physician, was deported to the ghetto of his hometown Lwów and later to the concentration camps of Auschwitz and Buchenwald. He survived, and after the war, resumed his work on immunobiology at the Polish Academy of Science before emigrating to Israel in the late 1950s, where he died in 1961. A thorough and ongoing discussion of his revolutionary ideas on thought styles and thought collectives, his "knowledge view" (*Wissensanschauung*) as he called it (1929, 430), only began at the end of the 1970s with the English translation of his book and its republication in Germany (1979; 1980). Fleck's long laboratory experience as an immunologist was the starting point for his deliberations on how communities of scientists in local working settings fabricate, develop, modify, and discard their scientific objects in their daily practice, how they gradually harden them into facts, which themselves are continually reshaped in an ongoing process. It is not single experiments, says Fleck, that are typical of the laboratory sciences. What is typical, rather, are extended series of experiments that communicate among each other in differing degrees and that constitute an experimental texture in which unforeseeable local reinforcements as well as unexpected eradications may occur. Where Fleck repeatedly compared scientific occurrences with a meandering river which, through its own movement in the interaction with the contingencies of the terrain, produces the very form and direction of its course, he was by no means of the opinion, to extend the analogy, that it would not obey the laws of gravity. He did not advocate a relativism of "anything goes." The resistance of the phenomena, and thus also their realness, remained constitutive for him. However, he did indeed think that gravity neither determines the particular form of the river of knowledge nor the precise point where it will join the sea—if indeed science has anything at all comparable to such a sea. He contended that its form and its actual end points were conditioned by what he called the "succession of the discoveries" themselves (Fleck, 1929, 429).

With Fleck's work, the perspective of an intrinsic historicity of knowledge is brought into play that has nothing to do with either the classical, if only asymptotic approximation of Riezler's second reality of scientific knowledge to the third one, the absolute and perennial outer world, or with a mere succession of contingent events, a merely accidental stream of discoveries, or a series of constructions directed by specific interests. On the contrary, history is now at work in the inner core of epistemology itself, in the *episteme* itself, which is conceived as a cultural configuration with its own temporality. The acquisition of knowledge, especially of scientific knowledge, becomes an iterative procedure, out of which the possibilities for the next round emerge only gradually and depend on the actual state of epistemic affairs. In this sense, Bachelard, whom we can regard in some sense as a French parallel to Fleck and who gives us the impression of a remarkable European dimension of this historicization of epistemology, also talks of a "rectifying pace," of a "thought in action" that is to be seen as the "veritable epistemological reality" (Bachelard, 1987, 300).

Again and again, Bachelard emphasized that the modern sciences are to be understood as "phenomenotechnologies," that is, as enterprises addressing and shaping their objects of investigation in the laboratory in a technical horizon (Rheinberger, 2005). As a rule, these objects are no longer accessible by means of everyday experience. In an inversion of the Cartesian philosophy of knowledge with its emphasis on the evidence of clear and distinct ideas, Bachelard regarded what were traditionally recognized as the most simple and basic scientific phenomena as being the most derived ones, because they were most subject to the phenomenotechnical work of purification. He urged his fellow philosophers and historians of science to pay close attention to this work of purification, to study it in great historical detail and to develop an epistemology of the detail accordingly. It is characteristic of Bachelard's epistemology of the detail that, under the concept of "epistemological obstacle," it regards a certain inertia and inevitable blurredness as being inherent in the very act of knowledge acquisition. Gaining knowledge means, in a certain sense, to outwit one's own current thinking in a phenomenotechnical act of surpassing. It finally allows one to arrive at facts that by definition are not accessible to the anticipating imagination, simply because they are beyond the reach of our everyday experience and our ordinary senses (Bachelard, 1969).

In his 1922 book on *Historicism and its Problems*, the well-known German theologian, philosopher, and politician Ernst Troeltsch diagnosed, as did so many others at the time, a crisis of the general philosophical foundations of historical thinking and of the cornerstones upon which and between which the nexus of history was to be "contemplated and to be construed." In the years after World War I, Troeltsch perceived, as he expressed himself, an "incredible longing for a concentration of historical life." He experienced nothing less than a foundational crisis of historicism, and he saw it as part of a general crisis concerning the sciences at large (Troeltsch, 1922a, 4–5). In his late years, Edmund Husserl argued structurally in a very similar manner. Between 1934 and 1937, that is until shortly before his death, Husserl worked on his book on *The Crisis of European Sciences and Transcendental Phenomenology*. In this book, he complained that the triumph of the positive sciences with their undeniable theoretical and practical successes and their orientation toward "prosperity" (he used that word in English) had paved the way for a pervasive "philosophical and *weltanschaulichen* positivism" that had become firmly entrenched and generally disseminated throughout society (*Crisis*, § 2–3). For a long time, the sciences had, according to Husserl, already become detached from the living world. They no longer had anything to say to people, certainly not to the young generation that opposed them with hostility. This development had led not only to a crisis of philosophy and the modern sciences in their philosophical universality, but also to a "crisis," as Husserl claimed, "of European humanity itself in respect to the total meaningfulness of its cultural life, its total 'Existenz'" (*Crisis*, § 5).

Husserl had drafted his *Crisis* when he was invited by the Viennese *Kulturbund* to present a lecture in Vienna, which he delivered in May 1935 (*The Vienna Lecture*). In this paper, Husserl outlined the program he then developed more systematically in his last grand opus. I wish to point out here that astonishing parallels emerge in that work to the program of Ludwik Fleck, despite the fact that the physician and microbiologist Fleck and the mathematically and logically trained philosopher Husserl could not be further apart as far as their academic socialization and their disciplinary backgrounds of experience are concerned.

Just as for Fleck, it was clear to Husserl that what was at stake was ultimately a new understanding and self-understanding of the *natural sciences*. And just as Fleck, Husserl realized that this new self-understanding could not be gained or recovered by treading the paths of traditional rationalistic philosophy of science. Instead, Husserl was determined to hold out against the usurpation of all understanding of science—and culture for that matter—by objectivism, naturalism, and positivism, by means of a countermovement; namely, one of bringing the academic disciplines back into the cultural historical horizon of living experience. Let us recall that Fleck had also questioned Riezler's distinction between a first reality of living experience and the second reality of scientific knowledge and even pleaded for a conflation of both.[1] At the very beginning of his exposition, Husserl asks: "Is it not absurd and circular to want to explain the historical event 'natural science' in a natural-scientific way, to explain it by bringing in natural science and its natural laws which, as spiritual accomplishment, themselves belong to the problem" (*The Vienna Lecture*, p. 273)? According to Husserl, such a "fatal naturalism" could only be counteracted by an extensive historical gesture seeking to understand the knowledge of nature as a cultural "spiritual accomplishment," as the historical product of the joint efforts of "collaborating scientists," Husserl's philosophical equivalent to Fleck's "thought collectives." With that, the knowledge of nature was to be brought from its positivistic residual existence and "residual concept," as Husserl put it, back into the horizon of a historically grown knowledge of the world (*Crisis*, § 3).

For Husserl, this in no way means relativizing the validity of scientific knowledge in the sense of a social or even merely epistemological constructivism. He wants to see it in flux. For him, just as for Fleck and Bachelard, science embodies "the idea of an infinity of tasks, of which at any time a finite number have been disposed of and are retained as persisting validities" (*The Vienna Lecture*, p. 278). Scientific questioning is a principally endless process without closure. One must see it as "mobile goings-on from acquisition to acquisition"—Fleck's "succession of discoveries," Bachelard's "knowledge in its course"—a process carried out by an "open chain of the generations of those who work for and with one another" (*Origin of Geometry*, p. 356). What is important for Husserl is that those working together and for each other are permanently aware of the fact that their questioning, in the last instance, is nourished by their living world and that even the oldest sedimentations of a scientific tradition, across all their transformations, remain ultimately bound by this living world. With a thoroughly programmatic gesture, Husserl asks and insists at the same time: "Where is that huge piece of method subjected to critique and clarification (–that method) that leads from the intuitively given surrounding world to the idealization of mathematics and to the interpretation of these idealizations as objective being?" (*The Vienna Lecture*, p. 295). In the eyes of Husserl, given that this critique and clarification remained undone, and that the self-critique and self-enlightenment of scientific workers, including the philosophers, remained undone, it was this deficit that was responsible for the crisis he diagnosed at a time when the national socialists in Germany celebrated their early triumphs and excesses, and against whom Husserl called upon the philosophers as the true "functionaries of humanity" (*Crisis*, § 7).

Husserl left us that "huge piece of method" for historical critique and clarification. In his short, untitled fragment on the *Origin of Geometry* posthumously published by Eugen Fink, Husserl sketched the contours of a historical epistemology that would devote itself to this critical task.[2] Here, Husserl rejected the supposition of a fundamental separation between theoretical *explanation* and historical *explication*. He pleaded for a combination of theoretical explanation and historical explication in an originary foundational effort and accomplishment. In this paper, Husserl stated with great emphasis: "Certainly theory of knowledge has never been seen as a peculiarly historical task. But this is precisely what we object to in the past. . . . What is fundamentally mistaken is the limitation through which precisely the deepest and most genuine problems of history are concealed" (*Origin of Geometry*, p. 370). Husserl struggled with these "deepest and essential problems of history." A "real, genuine history of the separate sciences," as he put it, could not simply be written as a recounting of their development in the contingency of past events. On the contrary, a true history of the scientific disciplines was for Husserl "nothing other than the tracing of the historical meaning structures given in the present, or their self-evidences, along the documented chain of historical back-references into the hidden dimension of the primal self-evidences which underlie them" (*Origin of Geometry*, pp. 372–73). Put simply, as Husserl paradoxically summarized elsewhere in the same paper, "what is historically primary in itself is our present" (*Origin of Geometry*, p. 373). In Husserl's view, borrowing the words of the French historian of science Georges Canguilhem, "the past of a science of today is not to be confounded with that science in its history" (Canguilhem, 1977, 15).

Husserl did not or could not follow the path he sketched to its very conclusion. His version of a historical epistemology essentially remained an attempt to catch up with foundational knowledge gestures. In the end, it remained an exercise of looking out for originary evidences he called the "historical a prioris" of a "teleological reason running throughout all historicity" (*Origin of Geometry*, p. 378). Husserl remained bound by such a teleological perspective, and in this point he certainly distinguished himself from Fleck as well as from the French tradition he inspired.[3] At the same time, however, he paved the way for later efforts to approach the peculiar historicity and materiality of the ideal objects of science from the ateleological perspective of a typology of forms of iteration. In *The Origin of Geometry*, we once again read with regard to the historical invention and intervention of the art of writing: "The important function of written, documenting linguistic expression is that it makes communication possible without immediate or mediated personal address; it is, so to speak, communication become virtual" (*Origin of Geometry*, p. 361). Husserl was inclined to see writing as a precondition for the peculiar historicity of science and the forms of sedimentation and iteration it made possible. It is the merit of Jacques Derrida to have insisted on this message of Husserl,[4] and thus to have made the late writings of Husserl available to a much more exoterically oriented historical epistemology (Derrida, 1999).

§ 11 Foucault, Cavaillès, and Husserl on the Historical Epistemology of the Sciences

David Hyder

Most readers of Foucault are eventually struck by a methodological conflict in his work: While Foucault is adamant that systems of knowledge are constituted within social contexts and maintained by power relations, he is equally convinced of the need for theoretical critiques, such as his own, of the history of knowledge. Such critiques are not intended to be positivist history; rather they have the unmistakably normative aim of freeing us of certain contingent and undesirable features of our (scientific) patrimony. In Anglo-American terms, Foucault walks like a social constructivist, but he often talks like a normative philosopher. By conventional measures, the methods of these schools are incompatible, and this is no doubt one reason that Anglo-American readers have tended to damn or to praise Foucault for his relativist leanings.[1] Either they overlook the normative aims of his work, or they simply ignore the tension between constructivist means and normative ends.

My purpose in this chapter is to explicate the sources of this seeming inconsistency by analyzing two key notions in Foucault's early work, by which I mean his writings up to and including *The Archaeology of Knowledge*. The first of these notions is the notion of archaeology itself, a form of historical investigation of knowledge that is distinguished from the mere history of ideas in part by its unearthing what Foucault calls "historical a prioris." This concept is the second of the two I will be concerned with. Such historical a prioris, on Foucault's account, are "unconscious" matrices governing the space of possible statements (*énoncés*) that occur in the writings of a given historical age. They are conceived simultaneously as conditions on the knowledge of a given scientific culture and as the framework that historians and philosophers use to classify the writings of the age they are studying. Both notions, I shall argue, are derived from Husserlian phenomenology. But both are modified by Foucault in the light of Jean Cavaillès's critique of Husserl's theory of science, according to which not only Husserl, but also most logicist philosophers of science and mathematics, are laboring under Kantian "transcendentalist" delusions.

According to Cavaillès, transcendental philosophies of science seek to justify logical and methodological norms by deriving them from foundational acts of consciousness. This approach is easily discerned in Kant: In the first *Critique*, the normative authority of logic and mathematics is both explained and justified by revealing their origins in those faculties of the mind whose operations are conditions for the possibility of unified and conscious experience. Drawing on his work on the history and philosophy of mathematics, Cavaillès argues that Kant and his successors, above all Husserl, are unable to account for the evolution of modern logic and mathematics on this scheme. Husserl, for instance, extends the transcendental program by focusing on the role of intentionality. Already in his earliest writings on mathematics and logic, he seeks to explain the normative character of these sciences by means of a phenomenological analysis of original intentional acts: Geometry, for instance, concerns "ideal" mathematical objects, such as triangles. These ideal objects can be the objects of intentions only to the extent that such intentions are systematically developed out of more quotidian ones concerning immediate spatio-temporal experiences. Such a process of idealization is not, however, fundamentally different from that involved in the objectification of other objects of experience. For instance, we also constitute a three-dimensional tree standing behind our immediate perceptions by imagining the tree viewed from multiple points of view. In both cases (the tree and the triangle), the objectification of the object goes hand in hand with certain constitutive operations of consciousness, which operations are the conditions on our establishing intentions regarding trees and triangles. A phenomenological investigation lays bare the normative content implicit in our intentional states by analyzing such primitive constitutive acts. Such acts are, as in Kant, the acts of a single, conscious subject.

I will not have occasion to go into Cavaillès's objections to logicism and the work of the Vienna Circle at any length in this chapter; however, it is worth touching on the parallels he sees between Husserl's and their work. Logicist theories of mathematics seek to reduce the content of mathematical propositions to logical truths. For both Frege and Wittgenstein, this analysis requires a further definition of these logical truths, and in both cases this definition presumes the existence of prior and structured *meanings*, for instance Frege's third realm of senses, or Wittgenstein's logical space. In order for the reduction to do any work, these structures must be assumed as given, for otherwise the scope of the phrase "logical truth" would not be well-defined, and thus the point of the reduction would be nullified. That is to say, if the principles of logic could not be identified a priori, then we might have to introduce supplementary logical principles as we moved along. And in such a case the claim that mathematics was in some sense reducible to logic would at the very least lose its persuasive force.[2] Traditional logicism, on Cavaillès's reading, founds the authority of logic by invoking the prior existence of meaningful symbols, and to its credit offers more than a merely formal justification of logical laws. Later versions, such as that of Carnap, abandon this presupposition; however, they do so at the cost of making logic part of mathematics—thus of abandoning the initial reductive project.[3]

Cavaillès is opposed to all these theories of science and mathematics and proposes in their stead what he calls a "philosophy of the concept," which is a philosophy that tries to understand the sciences by examining the history of concepts and the norms that govern their use. Such an approach is to be distinguished from its transcendental foils in that it does not demand or expect the closure that transcendentalism cannot do without. It does not, in other words, imagine that the ground of normative rules is to be found by revealing hidden cognitive structures, or what Husserl calls "sedimented" cognitive acts. Cavaillès complains that all such theories inevitably lead to the result that there cannot really be anything unexpected in science: The real objects with which science is concerned can never penetrate the fabric of cognition, because the latter must screen the real world precisely in order that it may the source of norms. Transcendentalism, on this view, is constitutionally opposed to realism, and it also cannot admit any real developments in scientific methodology.

It is, I shall be suggesting, just this critique of transcendental philosophy that provides the philosophical background to Foucault's early work. For example, in his contribution to an issue of the *Revue de Métaphysique et de Morale* dedicated to George Canguilhem's work, Foucault gives a brief summary of the state of French philosophy and history of science in the period after the war. In an oft-cited passage,[4] he suggests that there was a fundamental division cutting across the series of more easily recognizable cleavages between Marxists and anti-Marxists, Freudians and anti-Freudians, and so forth, namely, the opposition between a philosophy of the subject, of consciousness, and a philosophy of knowledge, "of the concept" as he puts it.[5] This division filtered the reception of Husserlian phenomenology in France, with the result that Husserl's thought was reworked on the subjectivist side by authors such as Sartre and Merleau-Ponty into the existential philosophy, whereas the conceptualists—among whom he includes Bachelard, Canguilhem, Cavaillès, and himself—were more interested in Husserl's early works on the philosophy of science and mathematics.

As I have just indicated, the term *philosophy of the concept* is introduced by Cavaillès during his critique of what he regarded as Husserl's inordinate appeal to the subject in the latter's final work, *The Crisis of European Sciences*. The research program that both Cavaillès and Foucault mean to describe with this term is, one might say, empirically rich but philosophically poor. Bachelard's writings on physics, Cavaillès's work on mathematics, Foucault's and Canguilhem's research on the life and human sciences offer detailed analyses of the emergence of new conceptual frameworks in the sciences; it is work whose interest is largely independent of its theoretical underpinnings. However, their theoretical assumptions are at best vaguely articulated, and their alternative philosophy of the concept is defined largely by negation: It is *not* a transcendental philosophy of science, in that it does *not* assign a central epistemological role to the conscious acts of subjects. Cavaillès is hostile to any project that seeks the sources of logic and evidence, method and data, in the minds of scientists. Similarly, when Foucault in the 1960s describes his work as concerned with the *un*conscious of science, he means to underscore what it is *not*, namely a phenomenological investigation of the history of the sciences. He is quite right to emphasize the distinction between his work and that of Husserl. But his need to do so stems from the overwhelming similarities between his theory of archaeology and Husserl's work. Indeed, both of our key concepts from the *Archaeology of Knowledge* (*archaeology* and *historical a priori*) are of Husserlian origin.[6] And both are repeatedly explained there in terms of what they are not: The discourses that comprise fields of knowledge and the rules and ontologies attached to them are, for instance, not conditions on the experience of historical actors. Of course, they do have the function of unifying fields of objects by means of rules, and in this sense their role is parallel to Kant's requirement that the understanding unify the data of conscious experience. But here again, Foucault denies that his theory is a Kantian one, claiming instead that his rules of discourse differ from Kantian rules of thought in being rules governing language. They are not "condition[s] of validity for judgments, but . . . condition[s] of reality for statements" (Foucault, 1969, 167). The historical a prioris constituted by these rules are not conditions on what could be *thought*, but on what could be *said* in a given science in a given age.

This insistence on the theoretical priority of language to thought is indeed characteristic of the French backlash against phenomenology in the period after the war, a backlash in which Foucault was only one of many participants. As I shall suggest below, the seeds for this reaction were planted by Husserl himself in his late recognition of what Merleau-Ponty called the "problem of language," that is to say, the problem of explaining how signs convey meaning in the absence of concomitant intentional acts. The division between Foucault and his structuralist allies on the one hand,[7] and phenomenologists such as Husserl, Merleau-Ponty and Sartre on the other, consists above all in the former's insistence that meaning should be analyzed by looking to the implicit structures of sign systems in preference to considering it as a cognitive phenomenon. This French "linguistic turn" is analogous to simultaneous developments in the English-speaking world, and it is characterized by a similar skepticism regarding the theoretical value of meanings or intentions, at least insofar as these are conceived as elements of consciousness.

As we just saw, in Cavaillès's critique it is not the problem of language that plays the central role, but rather the transcendental approach to the philosophy of science and logic. Nevertheless, the two topics are intimately connected, in that the problem of accounting for language and meaning within the framework of a transcendental philosophy is central to the late work of Husserl that Cavaillès criticizes.[8] And it is also, on his view, a central topic of early logicism.[9] The difficulty, in a nutshell, is that transcendental theories assume that the meanings of physical signs are mental events (senses, judgments, possible experiences). But it is equally clear that language has a capacity to generate, or at least convey, meanings that are not the products of our own conscious activity. Let me emphasize that this subordination of language to thought is not an accidental feature of such philosophies. It is, on the contrary, essential to their aim of explaining the binding character of normative principles. The reason that propositional logic is valid for linguistic statements, for example, is that it derives from constraints on propositional thought. Transcendental theories argue that in placing conditions on the structure of thoughts, these constraints also determine general connections among kinds of thoughts, for instance those holding among their categorial forms. These formal relations are codified in normative sciences such as logic, which tells us what sorts of transitions between sentences necessarily preserve truth.

This form of argument is therefore committed to the principle that meaning is determined by thought, and that language—the system of signs that refer to, or bear meanings—must conform to this determinacy requirement. Such a requirement is clearly exemplified by Frege's demand that every well-formed sentence of a logical calculus must have a sense: If this were not the case, then logical (syntactical) operations on other sentences could generate meaningless expressions, so that logic would not be truth-preserving. In other words, although logic is applied to *sentences*, it derives its normative force from meanings or thoughts. The problem, once again, is that it seems to be a simple fact about actual experience, and actual scientific practice, that we operate with meaning-bearing signs without any clear idea of their senses. We inherit both the vocabulary and grammar of the languages we speak, including those of formal scientific languages. And this fact puts pressure on the transcendental theorist: Since a speaker of these languages may never consciously have fixed their meanings, the theorist must explain where the meanings of such expressions are to be found, and such explanations run the risk of extravagance. Frege, once again, is obliged to postulate an entirely separate "realm" of senses, which are neither signs nor their referents. Husserl seeks refuge from this problem by arguing that the meanings of such signs are sedimented. At some time in the past, there were indeed conscious intentional acts that assigned meanings to these signs. If we want to know what the implicit rules of use governing these signs actually are, we have to inquire into their phenomenological roots. He concludes that it is an essential task of the epistemology of the sciences to identify and reactivate these sedimented meanings and rules.

The anti-phenomenological response to this problem of language is simply to drop the idea that meaning consists in hidden intentional acts and to accept the fact that sign systems acquire their meaning from elsewhere. One thereby shifts the investigation away from the cognitive acts of thinking subjects in favor of, for instance, social structures or speech acts. But it is evident that this solution carries a number of problems with it: Defenders of an intentionalist approach will rightly object that, whatever the legitimacy of social or behaviorist theories of meaning, they will never get at the very phenomenon that was to be analyzed. This objection is in fact programmatic, for both Frege's and Husserl's approaches were conceived as foils to naturalism: Frege's senses and Husserl's intentions cannot be construed as psychological without rendering the logical relations they determine purely contingent. In such a case, they would fail to be *Denknotwendigkeiten*, "necessities of thought," as Frege insists they must be. On my view, the negative definition of the philosophy of the concept that Cavaillès and Foucault adopt avoids the problem of language that emerged within phenomenology by denying the centrality of consciousness to meaning, but it does not adequately address the problems that remain once one has done so. The philosophers of the concept want the normative component of transcendentalist theories without the element of consciousness. This desire leads them to posit unconscious structural conditions on language, and these unconscious conditions are no less problematic than the hidden intentional acts they were intended to replace.

In the following, I will develop Cavaillès's criticisms of the transcendental philosophy of science by considering in sequence his critique of Husserl's late work, *The Crisis of European Sciences*, and then the criticisms of Kant with which he opens his *On Logic and the Theory of Science*. In conclusion, I will expand on the philosophy of the concept subscribed to by Bachelard, Canguilhem, Cavaillès, and Foucault and will consider above all the philosophical difficulties raised by their theory. My discussion will take the form of a close reading of two passages from the last page of Cavaillès's book, the first cited by Foucault in his essay on Canguilhem (Foucault, 1994b), the second cited by Canguilhem in an essay on Foucault (Canguilhem, 1967):

> Still, the justificatory evidence of transcendental analysis is necessarily unique: while there is consciousness of progress, there is no progress of

> consciousness. Whereas one of the essential problems of the theory of science is just that progress is not augmentation by juxtaposition, in which the earlier subsists with the new, but a perpetual revision of contents by extension and erasure.[10]

There is no one consciousness generating its products, nor simply immanent to them, rather it is each time in the immediacy of the idea . . . . Progress is either material, or between singular essences, and its motor is the need to exceed each of these. It is not a philosophy of consciousness but a philosophy of the concept that can provide a theory of science. The generative necessity is not one of an activity, but of a dialectic.[11]

Cavaillès's main objection here is quite simply that transcendental philosophy, at least in its traditional forms, can never explain how science can change. And the reason it cannot do so is that it is committed to the idea of an eternal and unchanging form of reason. Because the latter is, so to speak, already with us (in our minds, in our past, or in our language), the philosopher's task is inevitably directed toward the past. Both Canguilhem and Foucault endorse this analysis, because they are interested in analyzing the development of scientific rationality. In order to see what is at stake philosophically, I will unpack each of these quotations in turn.

The first passage I have cited is a *reductio*. Transcendental analysis seeks the normative roots of logic and mathematics in original (and in this sense constitutive) intentional acts. But this means that such analyses cannot make sense of a basic property of science, which is that it surpasses and obliterates its past. Therefore, such analyses are falsely premised: It is not true that there is only one ultimate domain of justification and evidence, namely human consciousness. As we have seen, this concern with the normative *origin* of scientific logic drives Husserl's phenomenological project. However, it only becomes an explicitly historical concern in his late work.

Husserl argues in the *Crisis* that we need a historical analysis of scientific concepts because these concepts are, in his terminology, sedimented, by which he means that they are concepts whose original meanings are concealed from us. As science progresses, it lays down layers of formalized systems, whose elements are increasingly remote from their experiential base. His primary example is that of geometry. Geometry was originally a technique of measurement immediately related to the life-world experiences and needs of the first "proto-geometers." In an initial, critical step, this knowledge was cast in axiomatic form. The formalization of geometry made it possible to reason concerning mathematical objects and to produce intersubjectively valid, eternal truths concerning these ideal entities. But this procedure severed the link between mathematical concepts and their life-world correlates. One could now think about geometry mechanically, by manipulating signs, so that one could do geometry without thinking about its phenomenal origins.

The next sedimentary layer Husserl considers in the *Crisis* is that of Galilean physics, which draws on axiomatized geometry without regard to its origins, ultimately positing it as the structure of physical space. It lays down a new layer of formalized science, namely the system of Newtonian mechanics, which also involves concepts with life-world correlates, such as that of force. These are also fixed symbolically, they sediment, and they are then reified as the world of mechanistic physics. Because this world is posited as the causal undercarriage of reality, it gives rise to a "crisis" in the sciences, for it now seems as though the phenomenal life-world, which is the true epistemological foundation of knowledge, is nothing but a supervenient illusion. The crisis can only be resolved, in Husserl's view, by reversing the process of sedimentation in order to reveal the genuine grounds of the sciences. It is this part of the phenomenological method that Fink characterizes as archaeology, a remark Cavaillès cites as an apt description of Husserl's project.

Thus Husserl thinks we need a historical epistemology of the sciences because the life-world is buried under a sediment of scientific concepts. By inquiring into the origins of these conceptual layers, we "reactivate" their original meanings, and thereby reveal both the source of scientific norms and the original source of meaning. Calling such an inquiry a true epistemology of the sciences, he anticipates the obvious objection. Why on earth do I need to ask about the conditions under which concepts were first developed in order to give account of their legitimate application? Surely I have the statements of geometry in front of me so that I can work back to their axioms and then inquire into their validity? Isn't that what it means to conduct an epistemological inquiry? Husserl responds to his own interjections by arguing that all concepts, indeed all cultural products, have an implicit historical dimension. We know that they were produced in the past, and we could not possibly make sense of questions concerning their legitimacy without having this fact in mind. For instance, to ask whether a cultural product such as a hammer is a good hammer is to know that it is a hammer and that it was produced by people to carry out certain tasks. These prior purposes are implicitly invoked whenever I consider the hammer as a tool, instead of as a lump of steel and wood. Similarly, I do not come to learn concepts, indeed I could not learn them, without first accepting them as cultural products with an intentional history, even if I do not know exactly what this history was. In order to apply them in new situations, I must at least *imagine* some past application.

What distinguishes this account of Husserl's from his earlier work is its insistence on the historical dimension, which is a late consequence of his desire to overcome Kant's formalism. Typically enough, Husserl's theory of scientific norms assigns a central role to intentions and their objects. In order to understand what our logic should look like, we inquire into intentional acts, which inquiry of course addresses their referents. Indeed, an essential aim of phenomenological analyses is to elucidate the *simultaneous* constitution of intentions and their objects. In doing so, we will come to understand what structures of consciousness are essential to our having intentions of a particular kind, for instance, mathematical ones. We are supposed to get a transcendental logic, but one that is founded on real experience, as opposed to intuitions and categories borrowed from the established sciences and posited as hidden cognitive faculties. Such an investigation therefore eschews aprioristic demonstrations in favor of phenomenological investigations of fields of concrete experience. In his earlier works, Husserl still imagines these constitutive intentional acts as occurring within the confines of a single consciousness. Just as in the related constitution-systems of Carnap, we are to imagine a single consciousness starting from a phenomenal base of givens in order to work its way up to the high-level concepts of scientific theory.

But by the time of the *Crisis*, Husserl sees this approach as overly simplistic. It is evident that the fields of ideal objects that make up scientific ontologies are not given in immediate experience—indeed they cannot be, because in order to have the degree of intersubjective validity and temporal invariance that we require from the sciences, their theoretical objects have to be distinct from the immediate experiences of given individuals. They are, Husserl emphasizes, first constituted as intentional objects when the appropriate written formalisms are introduced. But these formalisms are the products of a long cultural tradition. The reason that we need a *historical* investigation into the origin of geometry or into the origins of Galilean physics is that the formalisms that anchor the possibility of referring to ideal scientific objects are something that each individual consciousness has inherited. In order to know the logic that is appropriate to scientific objects, we have at least to imagine a situation in which they were constituted and sedimented, or in Husserlian terminology, the situation in which their ideal form was concretely available to a consciousness. Husserl recognized, in other words, that language could not be accounted for with his usual phenomenological methods, for the simple reason that the meanings of words are not something that the individual consciousness creates.

Because such a historical investigation must be undertaken from the point of view of the present, and because it is concerned with layers of sedimented meanings that "constituted" the objects of *past* societies and cultures, it cannot be carried out without postulating what Husserl calls a historical a priori (*Origin of Geometry*, pp. 377–78). Such a historical a priori might, for example, consist in the social structures and the measurement practices that formed the epistemological context in which axiomatic geometry emerged. This a priori is prior in a double, if not an equivocal sense: It is a prior condition on the cognitive actions of the proto-geometers; however, it is also an a priori condition on our conceiving of the emergence of their proto-science. This term makes clear that Husserl's history of the sciences cannot be understood as an empirical one. In writing a transcendental history of scientific concepts, we are aiming to discern paradigmatic formative events. Like the rational reconstructions of Lakatos, who was also inspired in this regard by Hegel, Husserl's reconstruction of the history of geometry doesn't try to get the facts straight. On the contrary, he would claim that there cannot really be any question of getting the historical facts straight until an analysis of his sort has been performed: If we are trying to imagine what went on in Greek society at the time geometry was created, we must obviously ascribe to the Greeks some sort of framework of meaning that we too can understand. There cannot be a question of reconstructing past intentions without assuming a conceptual framework common both to our forebears and to us. History, as we look at it, can be comprehensible only if we assume the existence of such schemes. At the same time, we regard our present scheme as the result of a historical process. These strata of meanings and traditions, which culminate in our own, form what Husserl calls the "internal structure" of history. A phenomenological investigation of the history of scientific concepts thus involves a curious fusion of empirical and a priori methods. It draws on historical data and a priori insights concerning human cognition in order to reconstruct the succession of conceptual schemes that produced the present one.[12]

So in Husserl's transcendental analysis, there is not only a search for the historical bases of current scientific concepts; there is also a search for their absolute normative ground. This ground is to be found in the primitive intentional acts of our imagined forebears. Thus Cavaillès's objection that, according to Husserl's theory of science, "if we are conscious of progress, still there is no progress of consciousness." We could not imagine the epistemological context of the emergence of geometry if we could not imagine what Husserl calls the "life-world" of other human beings. This life-world includes what some analytic philosophers today would call the world of subjective facts, the everyday world of colors and shapes, as well as the world of naive physics and psychology, for instance other people with their wishes and wants. Both our supposed Greek ancestors and we must inhabit such a world, whatever the current state of scientific development, for only on this assumption is the conceptual scheme of other human beings living in other cultures in any sense imaginable. It is, one might say, the base level historical a priori that will be brought to light by the archaeology of scientific concepts. But in uncovering conceptual strata and tracing the norms of science back to their phenomenal origins, we are engaged in a conscious *regress*. And to Cavaillès, this whole analysis is falsified by the true nature of scientific progress. Science does not consist of nested axiomatic systems; indeed, this is not even true of the history of mathematics, as Cavaillès argues in his work on logic and set theory. Earlier systems of science are not preserved as science develops; rather science develops by revising and erasing these systems, which may indeed be completely obliterated in the process. Epistemologically speaking, the conscious acts of our predecessors are irrelevant to us once we have rewritten their earlier system. It follows that the idea of a trans-temporal phenomenal ground of scientific knowledge is otiose, and with it the idea of a transcendental consciousness (or, more modestly, an internal structure of the history of reason) spanning the common human life-world.

Before turning to the second of the two quotations under consideration, let me summarize what we have learned so far. Transcendental philosophies of science seek to ground norms in the origins of conscious experience. They explain their validity by arguing that the latter are implicit in the structure of experience, whether this experience is my own or that of my predecessors. However, this philosophy is backward looking, and it cannot really account for the historical facts: Norms change, and ontologies are rejected and eventually forgotten. In the second passage cited above, Cavaillès suggests his alternative to this regressive philosophy. Instead of deriving it from the philosophy of consciousness, Cavaillès argues, we should conceive the doctrine of the sciences as being one of "concepts and dialectic." This remark invokes a specifically philosophical objection centering on Kant's transcendental deduction, and which Cavaillès develops in the opening pages of his book. This critique is, once again, extended to include Husserl's philosophy, but it is also intended to apply to the philosophy of science of the Vienna Circle. I will concentrate in the following pages on Kant and Husserl, because it is the distinction between "conceptual" and phenomenological theories of historical epistemology that is important to understanding Foucault's and Canguilhem's application of Cavaillès's ideas. I will then conclude with a few critical remarks concerning this application.

Cavaillès's main objection to Kant's transcendental arguments concerns the latter's attempt to justify the necessity of logic by invoking the unifying function of the categories for experience. This is why he insists on the fragmentation of consciousness, that consciousness is "lost in the immediacy of the idea" or that "the term consciousness does not have a univocal application." For, on Cavaillès's view, Kant cannot reconcile the generality of logic (Cavaillès sometimes speaks of its "absoluteness") with its transcendental origins. That is to say, Kant proves the validity of logic by invoking its constitutive role in experience; however, this form of deduction precludes the resultant logic's having general application. To see why this is so, one needs to recall why Kant's project requires a transcendental deduction of the categories, which are of course the basis of general (Aristotelian) logic.

If he is to justify the validity of this logic for all possible experiences, Kant cannot merely appeal to the fact that we *do* think by means of the categories. Even if we assume that they result from pure logical functions of the understanding, this does not explain why they are objectively valid, in other words why the understanding is always right to employ them.

Kant's solution is that these functions of the understanding are conditions for the unity of apperception. There would be no consciousness if these functions were not actively constituting our unified experience by connecting the data of intuition. In the language of the first *Critique*, to say that consciousness obtains is to say that there is a unification of a manifold in a series of cognitive acts. Kant's strategy is therefore to isolate the cognitive acts responsible for the unification of experience and then to extract logic by means of a dual process of abstraction.

Cavaillès, in his opening discussion of the role and the source of rules in science, characterizes Kant's deduction of these rules as follows: First, the empirical is reduced to the pure content of a mathematical form wholly determined by intuition. Second, the content of thought in general is sloughed off, leaving us with the pure functions of the understanding. Cavaillès objects to both these steps on the grounds that (1) we must be able to make sense of a content independent of its form in order to imagine its being eliminated, but then the form is not a condition on our conceiving the content; or (2) form and content are indeed indissolubly wed, but then the necessity inheres in this single composite, and we are not isolating the necessary condition by eliminating the part we call the matter or content. Thus Kant is wrong to think that he has deduced the validity, and thus the normative character of general logic, by means of his appeal to consciousness. The logic that governs judgments, if it can be shown to be valid, must involve an ontology, as does the logic of Aristotle's *Analytics*. And if it does not, then it is entirely empty. "In a philosophy of consciousness," Cavaillès concludes, "either logic is transcendental, or it does not exist."[13]

This attempted refutation of Kant, although highly compressed, does indeed strike at a notoriously weak link in the argument of the *Critique*. Cavaillès's point, in a nutshell, is that if Kant is consistent in maintaining that every cognition (*Erkenntnis*) contains a unifying concept and unified intuitions, then he cannot speak of a prior, non-unified multiplicity that is subsequently unified by the pure functions of the understanding. Conversely, and as Kant himself acknowledged, these pure functions and their associated categories are *empty* without the contents they unify. Kant's claim is that the categories, once they have been shown to be implicated at the origin of each cognition, have in consequence also been shown to be valid for *all possible subsequent* cognitions. This would indeed parallel Aristotle's (or at the very least the scholastic) justification of the logical role of the categories by means of their metaphysical one; in other words that because they are fundamental modes of being, they are also fundamental modes of propositions. But of course Kant doesn't want to do that: His categories are not modes of being, but modes of cognition. What they govern are all and only those experiences that can be "anticipated," to use Kant's language, namely the extensive and intensive magnitudes of which the multiplicity of phenomenal experience is composed.[14]

The problem, Cavaillès argues, is that this transcendental justification of logic has been falsified by our repeated encounters with *unanticipated* scientific objects that have contravened our logical principles. This flaw in Kant's system has been most recently revealed, in Cavaillès's opinion, in Hilbert's and Gödel's work on the foundations of mathematics. Hilbert's project of axiomatizing the various branches of mathematics was intended to provide these with logically secure foundations, although Hilbert did not seek to *reduce* mathematics to logic. He took the axioms to express synthetic, as opposed to analytic truths, and thus far he sided with Kant (and Poincaré) against the logicists.[15] But Hilbert, because he wished to establish the completeness and consistency of his systems, thereby put in play the logical rules involved. A system is complete when all truths are deducible within in it. Only in this case are the synthetic truths of mathematics strictly regimented by logical principles. Only then does the business of axiomatizing isolate the synthetic content of the science in question in the axioms, as Hilbert wanted. From this point of view, Gödel's result touches the heart of Kant's system, for it shows that there is no "general logic" adequate to mathematical knowledge, that is to the sciences of pure intuition. How much less should we expect there to be one for those concerning empirical intuitions? For Cavaillès, this result is the gravest symptom of a general ailment, namely that the Kantian philosophy of science "completely ignore[s] the contribution of the object to the structure of theory" (Cavaillès, 1960, 14). It does so because it demands that the essential structure of theory be determined by rational principles. These rational principles are in turn principles of the unity of consciousness, and thus the unity of science is nothing more than a projection of this unity onto the natural world. "There is no science in the sense of an autonomous reality that can be characterised as such, but rather rational unification following a fixed type of a diversity organized by the understanding" (14). There is no room for discovery and temporal contingency in this picture of science, just as there is no room allowed by the transcendental deduction for data beyond those that can be constructed or anticipated in pure forms of intuition. Both in the specific mathematical cases that Cavaillès dealt with in his earlier technical work, and in the general case of transcendental theories of science, the foundational character of logic can be preserved only by cutting its connection to the real complexity of scientific objects. Conversely, if logic is to retain its normative role in the sciences, we must concede that the latter cannot be deduced transcendentally.

Husserl, as we have seen, is taken to task on similar grounds: His search for normative origins is regressive, and it cannot make sense of an intervention in the world of science from outside of consciousness (indeed, the methodology of the *epoché* deliberately shuts out this possibility). Nevertheless, Cavaillès concedes that Husserl's work is superior to that of both Kant and his logicist descendants to the degree that it seeks the sources of logical and mathematical norms in concrete intentional acts. Indeed, Husserl, in the sections of the *Crisis* that were supposed to form its main body, and which were published only in the 1960s, characterizes his aims in a manner completely consistent with Cavaillès's take on the book, in other words with reference to the Transcendental Deduction. In these passages of Part III of the *Crisis*, Husserl argues that his life-world was implicitly assumed by Kant in the deduction, but that Kant failed to pursue the matter properly. Kant was right, claims Husserl, to begin the "descending" analysis of the A-Deduction in the world of everyday unified experience, but he passed too quickly over this "realm . . . of unexamined evidences of being" that are "constant presuppositions of scientific and philosophical thought" (*Crisis*, § 28). Kant acknowledged the existence of the life-world but failed to see that the deduction should be carried out on this level. In consequence, the logic and mathematics he extracted he had implanted artificially, that is to say Aristotelian logic plus the building blocks of Newtonian physics. Whereas, Husserl believes, the phenomenological method will yield the proper logic precisely because it is grounded in actual, as opposed to idealized, phenomenal experience.

You can imagine how Cavaillès responds to this: Husserl's method doesn't really get around the problems with Kant's. This "pseudo-temporal" origin of concepts is not only a fantasy, but even if it weren't, it wouldn't get around the dilemma that faced Kant already: Even if the original constitutive acts of consciousness force a given logic on us, why should, indeed how could, that logic be shown to be binding on experience at a later stage of science? Expanding on his critique of Kant, Cavaillès comments on Husserl that " if transcendental logic really does found logic, then there is no absolute logic . . . . If there is an absolute logic, it can only derive its authority from itself, and it is not transcendental" (1960, 66). Applied to the notion of a historical epistemology, this critique yields the result that such an enterprise is "an abdication of thought" because it looks backward to the necessities supposedly implied in foundational acts, instead of considering the open-ended future in which new objects and systems of concepts are developed. The reason we need a logic of the concept, or a dialectic, according to Cavaillès, is that logic is in fact always concerned with a range of possible experiences about which we know as yet nothing. Moreover, the logic we have in hand may always reveal itself inadequate to a domain we know well. This may appear to undermine the normative status of logic. But in fact it does so only if we are in the grip of the philosophy of consciousness. That is to say, only if we think that the validity of logic must be determined once and for all by the structure of cognition.

Despite the merits of Cavaillès's critique, it is difficult to discern the outlines of the philosophy of the concept that he, along with Bachelard and Canguilhem, intends to offer as an alternative. Within the philosophy of mathematics, Cavaillès favors Tarski's semantic approach over those of Carnap and Frege, arguing that Tarski did not fall prey to the formalist's desire to develop mathematics out of a single stock of privileged syntactic rules, but tried instead to discern the logic implicit in extant theories that are "rooted" in actual mathematical objects and practices.[16] In general, Cavaillès and his associates favor a philosophy of science that accords equal weight to the contributions of the objects of theory (which are conceived as independent of both the theories and the theoreticians concerned with them) to those of the theories themselves, and to those of the "critical rectifications" of working scientists. Their approach shares, in other words, many of the virtues and vices of recent analytic work in philosophy of science. On the positive side, the philosophers of the concept are pluralists and, to some extent, realists; they seek a theory of science that is true to the actual practice of science; and they retain an interest in normative questions. However, the theory that results is something of a grab bag, and it is better defined with reference to the formalist and transcendental theories that it rejects than it is through any individual philosophical or epistemological theses.

This is certainly true of Foucault's use of the term philosophy of the concept, as I argued at the opening of this paper. He uses this term in order to distinguish his own archaeological investigations from their most obvious forebear, namely the phenomenological archaeology called for by Husserl. Foucault's archaeology, as I will argue in conclusion, aims at uncovering the historical a prioris forming the deep structure of the sciences of various historical periods; however, these structures are more heterogeneous than those imagined by Husserl. But just like Cavaillès, Foucault does not want to throw the baby out with the bath water; he does not want his historical analyses to collapse into mere social history. This normative aspect of the French theorists I discuss in this chapter clearly distinguishes their work from recent English-language history and sociology of science, and it thereby reveals an essential difficulty inherent to any properly philosophical study of the history of the sciences.

Foucault's archaeological project up until his *Archaeology of Knowledge* can be characterized as a subversion of Husserl in the light of Cavaillès's critique. By this I mean that the overall structure of Husserl's scheme is preserved, while the key epistemological tenet is rejected. In an interview from this period, Foucault characterized his project negatively, as "trying to discover in the history of science something like the unconscious." The working hypothesis, he continues, "is that the history of science, the history of knowledge, does not simply obey the general law of the progress of reason, it is not human consciousness, it is not human reason that is in some sense the owner of the laws of its history" (Foucault, 1994a, 665–66). What does this scientific unconsciousness consist in? Well, among other things, it is a subliminal matrix that determines the identification and ordering of scientific objects, as is explained in *The Order of Things*. More generally, it corresponds to what Foucault calls in *The Archaeology of Knowledge* an historical a priori. Husserl's historical a priori was, as we saw, the internal structure of meaning that (1) conditioned the field of objects and methods that historical scientists, such as the proto-geometers, were concerned with; and that (2) enables *us* to understand their intentions and traditions. Foucault's historical a priori differs from Husserl's in that it does not describe the framework of past intentional acts. It is instead supposed to be the framework that made possible the formation of past *énoncés*, or statements, which are to be identified and analyzed without any reference to the conscious intentions of individual speakers.

Thus Foucault's a priori resembles that of Kant and Husserl in being an ontological ordering that is, for this very reason, a source of normative principles. That is why he calls it the source of the "rationality" of historical periods. But it is so only in his qualified sense, for it is not "human reason" but some other ordering principle that is at work in the sciences of historical ages. And it is not human thought but human speech that is normed. The most obvious examples of such historical a prioris in Foucault's work are the three *epistemes* that organize the scientific discourses of the Renaissance, the classical, and the modern ages in *The Order of Things*. These structures cut across disciplines, they are in large measure invisible to intentional historical actors, and yet they fundamentally determine the things that scientists in these periods do and say, the things they take to be significant, unworthy of notice, or indeed self-evident. In this regard, they resemble Wittgenstein's "grammatical" rules or Kuhn's related disciplinary matrices. In all three cases, one supposes that there is a historically contingent a priori at work, whose impact is not conscious to scientists or speakers all while it fundamentally determines their selections of problems and methods, as well as the sorts of ontologies they are willing to countenance. Archaeology, like phenomenology, seeks to understand the meaning of past science by revealing these historical a prioris.

Let me then conclude by elaborating on this critical distinction between Husserl's and Foucault's conceptions of the historical epistemology of the sciences. I will do so by citing a criticism that Canguilhem made of Kuhn in the Introduction to his *Idéologie et rationalité dans l'histoire des sciences de la vie* (1977) and which he took to show the superiority of the French "philosophy of the concept." Canguilhem concedes that Kuhn's notions of paradigms and of normal science are useful to the extent that they highlight the historical development of scientific reason. But, Canguilhem objects, they are too Kantian: By invoking the ideas of intentions and of regulating acts, Kuhn's theory allows these two to become unstuck. More seriously, since Kuhn "accords them an empirical mode of existence, as facts of culture" (Canguilhem, 1977, 23), what seemed to be a critical philosophy turns out to be social psychology. Canguilhem then goes on to contrast Kuhn to Bachelard and Cavaillès, for whom the rationality of mathematical physics inheres in the normative practices of mathematicians. The development of scientific norms must, on the view of Bachelard, Canguilhem, and Cavaillès, be sought within the actual practice of science, for "a science is a discourse that is normed by its critical rectification" (1977, 21). Canguilhem objects, in other words, to the same tactic that Cavaillès criticized in Kant, in which one considers norms (the pure functions of the understanding) in distinction from the thoughts or intuitions they necessarily govern. In so doing, one prepares the ground for the claim that, although empirical intuitions are not merely products of the understanding, they are nonetheless bound by its laws. But in Kant the necessity in question is a properly transcendental one. In Kuhn's work, Canguilhem objects: "The paradigm is a user's choice. The normal is what is shared . . . by a collective of specialists in an institution" (1977, 23). The norms in question are conceived naturalistically, and this means that there is no distinction to be made, from the point of view of the historical epistemologist himself, between rational and irrational science, except perhaps a pragmatic one.

On Canguilhem's hostile reading, Kuhn follows one of the two dead ends that present themselves once norms and contents come unstuck: His normal science is a social fact. The other path is the one followed by Husserl: Make the norms a product of a self-validating consciousness that interacts with scientific objects only after they have been processed by that very consciousness. This leads to a static and internalist view of scientific development: The autonomous contribution of objects and experiments collapses, while the norms become irrevocably fixed. Husserl's historical a priori, because it is ultimately a product of consciousness, will not permit real critical reflection, for such reflection is defined as a regression to life-world origins. And it rules out the possibility that changes in norms be provoked by unexpected inputs on the side of the objects themselves. This is why Foucault, while following Husserl in identifying aprioristic elements in past science, nonetheless feels compelled to situate them on a plane that will prevent such a collapse. It is on this plane, he claims, "that a culture, stripping itself of the empirical orders prescribed to it by its primary codes, stops being guided by them passively . . . and liberates itself enough to determine that they are perhaps not the only ones possible, nor even the best ones" (Foucault, 1966, 12). This plane of ordering is neither that of Kuhn's social conventions, nor is it that of a Husserlian historical a priori, which is a structure of consciousness.

Of course, one cannot help but suspect that Foucault's solution to Canguilhem's and Cavaillès's dilemma comes at a high price, for we would certainly like to know where this plane of unconscious orders should be taken to lie, in other words what the substrate of the scientific unconscious is supposed to be. Furthermore, if this a priori is neither a structure of consciousness nor a sociobiological entity, how does it become actual? That is to say, in what sense does it cause or constrain the things that scientists say? If this a priori acts causally, then it provides an explanation of certain regularities that we observe in past scientific cultures. But then it would be what Canguilhem dismissed as mere social psychology. On the other hand, the demand that it be unconscious means that the historical actors are themselves unaware of it. If it is a system of rules governing their speech and conduct, then these rules are not something that they deliberately instituted and willfully follow. They are at best, like the basic rules governing Wittgensteinian forms of life, rules they follow blindly. And the most a historical epistemologist can say is that the authors of this period act *as if* they were following rules of this sort. No one can deny him his right to interpret historical events on this pattern. But if we are to speak here of a historical a priori, then we can mean by this at most the a priori of the historian. An essential feature of this notion in Husserl—that the historical a priori was binding on the historical actors—has gone missing. With it has gone the possibility of any properly normative stance toward these past scientific cultures. And that is just the criticism that Canguilhem levels at Kuhn.

My conclusion is that Foucault's theory of science, at least up until the point that he wrote *Archeology of Knowledge*, was largely defined through his *rejection* of Husserl's model, a rejection that he and others found in their reading of Cavaillès. And though that rejection is well grounded in the critique of transcendental logic and of the philosophy of consciousness that we find there, it retains too much of Husserl's historical epistemology for it to be coherent. For the notion of an unconscious a priori deliberately preserves features of the transcendental philosophies it was supposed to help supplant. These features—the notion of an a priori, and of the simultaneous constitution of logic, rationality, and a field of objects—are located, as Foucault explicitly maintains, neither on the level of facts, nor on that of consciousness. If they are conditions of knowledge, the knowledge they condition is most likely our own. That this terminology is absent from Foucault's later work is no doubt a reflection of his own misgivings in this regard.[17]

§ 12 Concepts, Facts, and Sedimentation in Experimental Science

Friedrich Steinle

My interest in Husserl centers on his notion of *sedimentation*, a notion that has proved particularly useful in my work in the history of science. Rather than offering a close interpretation of his writings, I shall approach Husserl in this chapter "from outside," by presenting a specific case from the history of science. I shall consider some notions that appear to be appropriate to that case and relate them to Husserl's concept of sedimentation. In the last section, I shall sketch the more general picture of scientific development that emerges from cases like the one under scrutiny and point out how the notion of sedimentation could be useful here.

Concepts: From Research to Language

Let me start with some observations of everyday and academic life. On the science pages of newspapers, we often find reports from specific research fields, addressed to a broad audience, for instance, the following:

- Certain populations, or subpopulations, have a higher risk of contracting AIDS, and we know already that AIDS is an infectious disease, caused by a virus, HIV.
- Modern versions of high-speed trains, such as the Eurostar, enable uninterrupted traffic over national borders by using modern thyristors, elements that make different alternating currents unipolar for high-power requirements.

- Insights into the dynamics and kinetics of chemical reactions of carbohydrates enable the planned design of plastic materials, pigments, and pharmaceuticals.

These reports present modern achievements of science, and in doing so, they tacitly use numerous scientific concepts: concepts such as infectious disease, virus, population, alternating current, bipolarity of electricity, chemical reaction, compound, and pigment. In itself, this is nothing exciting—there is no way it could be otherwise: We cannot express anything without concepts, and the specialization of modern science and technology leads to specialized concepts. But that is not all there is to it. All of the concepts I have highlighted were once created and formed in a specific context of intense scientific research, and from there found their way into common language. At the site of their origin, they were open, tentative, instable, and flexible, while later on they appear as solidified, stable, either as "natural" categories, or in some cases just as "facts." There is a long path from the openness of scientific research to the stability of concepts that are regarded as expressing simple facts.

This path, this development from openness to the closure of concepts has not yet received the attention it deserves. For much recent epistemological debate about such concepts and their features is conducted mainly in an analytic spirit, and there is little reference to historical or actual cases of how concepts come into being.[1] As a consequence, an essential feature of concepts is not taken into consideration: the process of their formation or, in other words, their successive development from tentative proposals to fixed elements of language; that is to say, there is little regard for a possible historical dimension of concepts. But this neglect leaves essential points out of the picture, as can be illustrated by my examples above. Can we fully understand the concept of chemical reaction, for example, without considering its historical development and, in particular, the alternative conceptualizations against which it finally was favored and accepted? The path from scientific research to scientific language or sometimes even everyday language is of essential importance both to our understanding of the epistemological features of concepts and to our understanding of science and the validity of its conceptual foundation. As I understand Husserl, it was exactly this aspect he was interested in when he thought about the origin of geometry. And it is just these processes of conceptual stabilization, solidification, and, of course, sedimentation that I shall focus on in this chapter. However, whereas Husserl was dealing with geometry, my interest is in empirical or even experimental science. I shall develop these topics by examining a historical case, following the pathway of a specific concept from the complexities of its inception, via its first stabilization, to the sedimented state in which we use it today. From this case, significant perspectives open up concerning the details and meaning of processes of sedimentation, the relation of concepts and facts, and our general picture of empirical science and its development.

Charles Dufay and the Two Electricities

One of the most basic notions we associate with electricity today is its bipolarity. Electricity is positive or negative—this notion, or even fact, is so basic for us that it is usually given no extra emphasis or explication but is merely used as part of the very language employed in using electricity and explaining its use to others. Indeed, it is often taken to be a defining feature of what we call *electricity.* Plus and minus, positive and negative are fundamental categories we refer to when we plug in a new battery in our camera or when we need our neighbor's help to start our old car after a frosty night. This has long been the case, at least since the mid-eighteenth century (when electricity meant what we call now static electricity).[2] Twenty years earlier, however, in 1730, the field was totally different. There was no thought whatsoever of two electricities. The concept was not proposed until 1734, but within slightly more than a decade, it led to a fundamental shift on the conceptual level. It is this shift that I shall focus on in the following.

The main actor is Charles Dufay, a brilliant early eighteenth-century academician, director of the Paris botanic garden, and versed in diverse domains such as botany and astronomy, dyestuffs and fire pumps, fluid dynamics and magnetism. The *Jardin Royal* offered ample resources for such experimental research. Dufay had, in an extended experimental series on shining minerals (Bolognese stones, as they were called), shown his interest in treating research fields in which even basic ordering was missing, and it was partly this interest that made him take up research in electricity. In the 1730s, the field was in an unstable and incoherent state.[3]

More than a hundred years of research throughout Europe had produced a multitude of different and puzzling phenomena, for instance:

- that some materials could be electrified by rubbing, others sometimes, and others not at all;
- that sometimes electricity acted as attraction, sometimes as repulsion; and
- that sometimes sudden changes in attractive and repulsive effects occurred.

Dealing with those questions proved difficult, even more so since the experiments were delicate, the effects tiny, and reproducibility was difficult. Dufay conducted extensive experiments, varying his procedure in many ways. He used a vast number of different materials, in individual or combined arrangements, and varied their shape, temperature, color, moisture, air pressure, and the experimental setting: two bodies in contact, in close proximity, at a large distance, connected by a third, and so on. His work led to remarkable results and bold claims, such as that *all* materials except metals could be electrified by rubbing, and *all* bodies except a flame could receive electricity by communication.

But still he was left with serious questions about when attraction and repulsion occurred and when the one sometimes suddenly switched into the other. He devoted extensive effort to this point. In a first step, he intended to clarify whether or not repulsion existed—Otto von Guericke's report of the effect (1672) had been contested by some. By varying his experimental setups, he analyzed the conditions under which repulsion occurred. He succeeded in obtaining repulsion even when the arrangement of other bodies in the vicinity was changed drastically; thus the repulsive effect was clearly shown to exist. In a particularly delicate arrangement, Dufay was able to keep a very light leaf of gold hovering at a distance of more than one foot over an electrified glass tube for several minutes, even when he moved through the room with the tube—much after the model of Guericke, who had described and depicted a similar experiment (Guericke, 1672). When the force of the tube lessened, the leaf would lower and finally fall down to the floor.

Dufay was thereby confronted with the question of when exactly repulsion occurred in contrast to the usual attraction—after all, attraction counted as the defining feature of the electric virtue. Again, he approached the question experimentally, by varying many parameters of the arrangement:

- the manner of electrification (rubbing, communication)
- the degree of electrification
- the size of the electrified body
- the material of the electrified body
- the basis/support of the electrified body (more or less idio-electric material)
- the arrangement of the bodies involved (larger or smaller distance)

For a long time, the results were merely puzzling, appeared instable, and did not resolve into anything resembling a regularity or correlation. Dufay could not formulate the conditions of when attraction and repulsion occurred, let alone when, as sometimes observed, a sudden switch between the two actions took place. This latter effect, however, was ultimately the first to be conceptualized in a rule. For specific arrangements Dufay was able to formulate regularities:

- Electrified bodies always *attracted* small unelectrified bodies (this was nothing new, but was just reiterated to clarify the contrast to the following statement).
- However, when these small bodies approached the electric one and themselves became electrified by communication, the former attraction turned into *repulsion.*

Dufay found this regularity, this law about the temporal sequence of attraction-contact-repulsion, to be valid without exception, and it could account for many of the effects he had already obtained. Nevertheless, it was restricted to the interaction of any pair of bodies in which the first had been electrified by communication from the second. In other cases, the regularity did not hold; they still appeared irregular. But Dufay did not stop here. He found a crucial hint in an experiment with a hovering gold leaf. When he approached a third electrified body to this leaf, strange effects ensued. If the third body was of glass, it repelled the leaf strongly, as the tube did. But when it was of copal, it was strongly attracted. Though this result was most puzzling,[4] it gave an indication that the material used was crucial.

In pursuing this hint, Dufay became increasingly aware of regular behavior that could not, however, be formulated in the usual language of electricity. Sensitized by his former work, particularly in botany and luminescence, he was more aware than other researchers of how essentially such work depended on working with appropriate concepts and categories. More than others, he was ready to question and revise even the fundamental categories of the field, and this is what he indeed did in a radical way: In order to express the regularities, he proposed that, rather than speaking of electricity in general, one should more accurately speak of two electricities. The regularity then was that electrified bodies repelled all those that had the same electricity, but attracted all those that had the other. As the experiments showed, the electricities maintained their character even when communicated to other bodies. Thus the above regularity of repulsion turned out to be just a special case of the now general regularity. According to Dufay, which of the two electricities a body acquired on being rubbed depended only on its material. Thus the dichotomy of electricities induced a division of all materials into two classes. At the same time, the electricities could be characterized according to these classes—Dufay named them vitreous and resinous electricities, respectively, in reference to particular prominent representatives of the two classes of materials.

We have no explicit notes about what exactly convinced him of the power of these new concepts. Presumably their immediate "success" was central here: Dufay emphasized that, with these concepts and the regularities they allowed him to formulate, he could understand not only his own, quite numerous experiments but also those reported by others. The concepts enabled him to do exactly what previously had been impossible: to order the whole field in such a way that stable regularities could be formulated. This was indeed a fundamental achievement, and Dufay was well aware of that. He not only announced his proposal immediately at the Paris Academy (as he did usually), but he also took the unusual step of writing a letter to the Royal Society in London to make his proposal known (Dufay, 1734). This striking deviation from his usual method of communication indicates how well he was aware of the fundamental character and importance of his new proposal.

How the Two Electricities Were Received

It was a very specific level of knowledge that Dufay aspired to and achieved. He was well acquainted with the various theories that attempted to explain electric effects by air currents, by humorous components of matter, by specific electric effluvia, and so on.[5] However, those theories did not appear in his research. His aim was not to consider the hidden causes of electric effects but to order them on the phenomenological level, to establish regularities, and to find the appropriate concepts that would allow him to do so. It would mean "attempting the impossible," he emphasized, if one were to search for causes without previously having discovered the large number of phenomena and having ordered them along a "few simple principles" (Dufay, 1733b, 476). From what Dufay actually did, I interpret this as the task of establishing regularities and laws. And in such a context, there was no space for theories about hidden causes. The search for correlations and regular dependencies dominated his enterprise, and in the end he was successful. But his success depended on a step that he had not planned or envisaged at all: on the introduction not of a new theory but of something more fundamental, namely, a new conceptualization, a new way of speaking about and manipulating electricity.

Before Dufay, there was only an electric virtue, defined by the attractive action onto small bodies nearby. With Dufay's innovation, the field looked much different. There were two species of electricity now, defined not only by their action on small bodies nearby, but essentially by a mutual interaction that could be either attraction or repulsion. Moreover, that dichotomy induced a classification of all materials according to their susceptibility to those two electricities when subjected to friction. In handling electrical effects, in designing, conducting, and evaluating experiments, new considerations became prominent. New questions were required and thus arose, questions that previously had simply been inconceivable because the very categories they used did not exist. For everyone who would adopt the new conceptualization—for the time being this was just Dufay himself—the field of electricity appeared totally transformed.

It is striking to see how the other electricians of his time reacted to Dufay's proposal. The first historical observation is the nearly complete absence of an explicit discussion of the topic. This is all the more significant, as the proposal did not disappear. On the contrary, it soon showed up again, but in a very specific manner. Only five years later, the writer of a most important physics textbook of the period, Leyden professor Pieter van Musschenbroek presented the new concept as unproblematic. "Experience has taught us that electrical power is of a twofold manner . . . ," he told the reader.[6] It is worthwhile noting that he did not say "*Dufay* has made this claim," but "*Experience* has taught us." Of course, this was not a denigration of the effort and achievement of Dufay. It was a significant indication of the specific level of knowledge that was addressed here. The existence of the two electricities was presented as indisputable, as a fact of nature, as we would say—and indeed, this quickly became the common view.[7] Five years later, in the first textbook ever devoted exclusively to electricity, Petersburg academician Johann Gabriel Doppelmayr went even further and took the twofold nature of electricity as a very part of its definition (Doppelmayr, 1744, 1). And this was characteristic of a general attitude. In textbooks of the eighteenth century, the twofold nature of electricity was very often introduced as a defining feature, derived "from experience," and only later in the text were the specific laws of attraction and repulsion presented (Frercks, 2004). Already here the concept was detached from the context in which it had been framed and was presented as an ontological feature, as a fact that as such was no longer subject to dispute. Such a way of "reception," of "filtering in," is strikingly different from the way in which theories proper are received, both by the absence of explicit discussion and by the quick introduction of the concept not as something to be problematized, but as a fact, taught by experience.

Sedimentation and the Impact of Concepts

As to the background of such a peculiar and striking way of receiving a proposal, I suggest that its peculiar character is due to the specific epistemic level in question here: It is not theories in some strict sense that are at stake in such cases, but the concepts themselves. Concepts and categories, as elements of language, cannot be true or false but are rather appropriate or inappropriate, useful or useless. They cannot be proved, confirmed, or disconfirmed, but they have to prove themselves against certain goals. What exactly the goals are, what exactly should count as "appropriate" means, is a matter of the specific historical situation. For Dufay himself, the touchstone was that any new conceptualization had to permit what the former had failed to achieve: the formulation of a coherent and general law of attraction and repulsion.

There may be other criteria, and for many others there have indeed been other criteria, though closely related ones. Perhaps the most central one had been success in manipulation. The new concepts allowed for an unprecedented stability in practically dealing with electrical effects. In 1744, Wittenberg professor Georg M. Bose redesigned the older electrical machine in a way that made a still greater stability of electrical experiments possible, and this provided the major background for the explosive spread of electrical performances throughout Europe within a short period. It is remarkable, however, that Bose explicitly made reference to Dufay and his two electricities (Bose, 1744). When, in 1746, the news of the Leyden jar shocked electricians throughout Europe and marked another starting point for what is sometimes called the "golden age" of electricity (with strong electric machines, spectacular displays at public events, and the first commercial uses of electric effects), the discourse of two electricities had become unproblematic and even "naturalized." Students of electricity took it straight from their textbooks; users of electrical machines learned of it in the workshops where they bought the machines.

This overwhelming instrumental success is one of the most intriguing features of the golden age of electricity. It can be clearly seen now, however—and this has long gone unnoticed in the histories of electricity—that the foregoing reconceptualization of the field provided one of the essential conditions for this success. Reconceptualization may have far-reaching consequences, as this case illustrates. The new concepts were, in a sense, incorporated into the machines and instruments, for example via the choice of materials and the design of the apparatus. At the same time, it was exactly this success in dealing with electricity, in inventing and using new instruments, that provided a central support for the appropriateness of these new concepts.

Once again, the different epistemic levels involved here become visible: In contrast to the formation and quick stabilization of the *concept* of two electricities, the situation of *explanatory theories* remained incoherent and unsolved, and this state would not improve with Franklin's theory of the 1750s. The debate over whether one should explain the bipolarity of electricity (to use a later name for what Dufay had called two electricities) by assuming one or rather two electric fluids would last for many decades, to remain unresolved and to run dry toward the end of the century. However, the very concept—or fact—of bipolarity was never questioned in this debate. It was not any clarification on the level of explanatory theories that furthered the spread of electricity since the mid-century, but the lasting success of the new conceptualization achieved by Dufay.

What we see in this historical case is how a new concept may be created in a quite specific context, by an author with definite, if not idiosyncratic interests, and in the course of broad and systematic experimental exploration.[8] We see then how this concept has been stabilized and has found, in a rather brief period of time, its way into the very language of the research field. Such a process recalls what Husserl describes as the process of sedimentation, a process that results in the end in the existence of a concept that we use in ignorance of its original "meaning."

Before I examine this analogy more closely, let me add that the case I have presented is far from unique in the history of the experimental sciences. Such processes occur again and again, in fields of research as diverse as chemistry, physiology, physics, and genetics, and in quite different periods, from the early modern period until the present day. Just to mention a few cases from the history of electricity, one may take the concept of a current circuit, covering the battery and its external wire at the same time and describing these two with the same language, a concept that Ampère framed to formulate a law of electromagnetic action (Steinle, 2002). Or one may think of the concept of magnetic and electric lines of force that Faraday developed in the course of twenty years of intense experimental research, and that allowed him finally to describe a vast realm of effects with very few laws (Steinle, 1996; 2005). In a different field, consider the concept of chemical reaction that was shaped and developed in the seventeenth century on the basis of broad experimentation, not by single prominent individual researchers, but rather within a community with a tight communication structure (Klein, 1994). Lastly, we may take the concept of infectious disease, a case to which I shall come back in a moment. Different as these cases may be, they illustrate well the fundamental character and the far-reaching impact of such processes of concept formation and stabilization.

Concepts and Facts

While these examples of generation, stabilization, and solidification of concepts resonate strongly with what Husserl calls sedimentation when discussing the case of geometry, there are also significant differences. A first indicator is the historical actor's language. While geometricians are well aware that they deal with (man-made) concepts, in the empirical and experimental sciences the language is different: One speaks of facts (usually in contrast to theories), by which one means states of things as they are in nature, states that are not inventions of the human mind, and that are taught directly "by experience," as Musschenbroek put it. In contrast to geometry, moreover, the awareness that basic concepts were once actively formed and stabilized tends to disappear in scientists' own accounts and in our usual understanding of facts. Facts are taken as a given, and thus as unproblematic and undisputable starting points of theorizing. Even in philosophy of science, there is sometimes little awareness that facts might not just be given, but necessarily involve conceptual (not necessarily theoretical) activity, effort, and even creativity.[9] A closer look at the practice of scientific research provides us with a more complex and differentiated picture, both historically and epistemologically.

Such alternative accounts were provided early on. Strikingly enough, it was roughly at the same time that Husserl wrote his *Origin of Geometry*, though independently of Husserl, that the young and unknown Polish immunologist Ludwik Fleck published a book whose very title was a provocation to many: *Entstehung und Entwicklung einer wissenschaftlichen Tatsache* (*Origin and development of a scientific fact*) (Fleck, 1935; 1980). It is still an intriguing historical question as to why exactly so many deep reflections on science were produced in the 1930s, for instance Popper's 1934 *Logik der Forschung*, Fleck's 1935 *Entstehung*, and Husserl's 1936 *Ursprung der Geometrie*. One common background was certainly the reaction to logical empiricism, but there must be more to say. The three accounts differed drastically, both in their direction and their fate. Fleck's book, in particular, written in German by a Jewish author, found no resonance in the political atmosphere of Nazi Germany. Even later, not the least through the dominance of the Logical Empiricist immigrants in the English-speaking world, there was no interest in an English translation. It was only Thomas Kuhn in the 1960s who realized what Fleck had to say, but it would still take until 1979 before the book was available in our modern lingua franca.

Among the three, Fleck's approach was the only one directed primarily at experimental science, in particular at his firsthand experimental experience as an immunologist. He presents in great detail the path that led, in immunology, to identifying the agent of syphilis and to the proposing of a reliable test, a development that took place at a Kaiser-Wilhelm-Institute in Berlin, and for which its main actor, August von Wassermann (1866–1925), received worldwide honors. Fleck shows in detail how much this research was based on concepts such as infection, disease, and immunity—notions that were taken for granted by many, but which, on closer inspection, bore significant traces of their genesis, and in particular of the decisions made within that development. Fleck emphasizes how much research depends on the existence and use of concepts that are taken for granted: "Every system will then become self-evident know-how itself. We will no longer be aware of its application and its effect" (Fleck, 1980, 114).[10] In order to grasp this phenomenon, he proposed his well-known notions of *Denkgewohnheit* and *Denkstil* (habit of thought and thought style).

Moreover, Fleck pointed out that when concepts disappear from the realm of explicit discussion, they are often turned into facts, that is, into a category that *per se* is taken as being immune to theoretical objections. Facts, in the usual understanding, present things just how they are, and we learn them from experience. But we have to use concepts and, as Fleck shows in detail from his historical case, these concepts are not just given but have been created, formed, and stabilized in a complex process, a process that lies in the past and is usually not recovered. In empirical sciences, the generation and stabilization of concepts is closely related to the genesis of facts. This is why Fleck, in his very title, speaks about the genesis, not the discovery, of facts.

As a side remark, let me note that Fleck, while he repeatedly emphasizes that these processes are inherently bound to social constellations and to communication, likewise puts emphasis on experimental outcomes and theoretical stringency. He is sometimes read as propagating a purely social-constructionist view of science, as explaining scientific development by changing sociological constellations alone, but this is clearly a misinterpretation. On the contrary, what he did was put his finger on the fact that there is no simple account of science, one focused on theory, experiment, or social setting alone, but that we have to develop a multifaceted view. It is this point that Kuhn took up and elaborated.

Sedimentation: Geometry and the Sciences

Fleck's historical case shows striking similarities to the case of the formation and stabilization of the concept of two electricities. Both of them illustrate a process in which concepts are formed and developed in an open way and later become fixed to such a degree as to form stable and unquestioned elements of the language of the research field. It is not only a process of stabilization we see here, but also of a sort of sedimentation: The now stable elements serve as an unshakable foundation, as the unquestioned ground for further research. Of course, processes of such a type remind us immediately of Husserl's discussion of sedimentation.

Husserl introduced this notion in discussing the origin and development of geometry.[11] By attributing fixed notions or symbolic signs to geometric concepts, he says, their original meaning may disappear in the course of time. Later generations take the signs and receive them passively but no longer reactivate their original connotations and meaning. Indeed, such a process is inevitable in constructing the "*Stufenbau*," the ever-growing system of geometry: Without it, everyone would have to start from the beginning and would waste her or his whole power in taking the first steps again. There is, in principle, always the possibility of "reactivating" the original meaning of geometrical concepts. As a matter of fact, however, this is simply not done. Since the early modern period at least, geometry has been in constant danger of becoming a "tradition empty of meaning," such that "we could never even know whether geometry had or ever did have a genuine meaning, one that could really be 'cashed in'" (*Origin of Geometry*, p. 366).

While Husserl articulates all this with respect to geometry and its development, he seems to see a fundamental contrast to the empirical sciences. There are only rare, but telling references to the sciences in his text. For the "so-called descriptive sciences," his analysis looks totally different: At least in principle, he says, every new sentence can be "'cashed in' for self-evidence" (*Origin of Geometry*, p. 363). The basic idea seems to be that there is no such Stufenbau as in geometry, where the original meaning of a sentence can be veiled, but instead always a direct access to that original meaning by empirical evidence. Empirical evidence enables and guarantees, so Husserl seems to say, the authenticity and directness of our statements about the world. As a consequence, neither a process like sedimentation nor the problems connected with it play a role in empirical science.

To be sure, this is a strong interpretation, based on the few textual passages he gives in his text. And it is not clear to what extent Husserl would extend the compass of the descriptive sciences. Did he just think of natural history, or would he also embrace fields with instrument-mediated observation and even experimentation in his analysis? The text does not focus on these questions and does not provide enough material to decide them. But if we regard Husserl's statement about descriptive science and the immediate evidence it provides as a general statement about observation, we might well take it, for a moment, as a statement about empirical science in general, as contrasted to the "deductive sciences" exemplified by geometry (*Origin of Geometry*, p. 365).[12]

Such a picture of empirical science would contrast strikingly with my reflections above concerning (empirical) concepts and facts. Rather than always recurring to direct and immediate evidence, these considerations tell us, empirical science has recourse to facts—but these facts are, no less than the concepts of geometry, the result of successive historical work, of processes of formation, stabilization, and *sedimentation*. And again, as in geometry, the main medium of sedimentation is language. The "seduction of language" exists no less and is no less effective in the sciences than in geometry (*Origin of Geometry*, p. 362). The very concepts that scientists use to express their empirical results can be subjected to an analysis similar to that developed by Husserl for geometrical concepts. Of course, Husserl's treatment of Galilean science in the *Crisis* does offer the outline of such an analysis. However, even here Husserl discusses the mathematization of the world as if it were a purely philosophical or theoretical activity, and not one that transpired in large measure in the scientist's workroom.[13] What is more, he still has only those sciences in mind that have been mathematized, as if the process of sedimentation was restricted to mathematical concepts. He does not envisage the possibility that those processes might also occur in nonmathematical fields, that is, in large parts of empirical science, even in fields that he calls descriptive sciences. Cases like Fleck's, or the one presented above, indicate that sedimentation is significantly more endemic than Husserl envisaged.

The Picture of Science

However, there is a disquieting observation: All these notions of sedimentation and solidification point to a cumulative picture of scientific development. Indeed, Husserl is often read as adhering to a picture of a science formed by successive layers, such as we have in geology, which accumulate in solid formations. At the same time, such a picture of science has been forcefully attacked by Kuhn, and with good reason. When we discuss and emphasize these processes of sedimentation, does it follow that we have to give up Kuhn's insights, and all the others that have followed, in order to return to the older, cumulative picture?

I would argue that this is not the case. I will, once again, illustrate this with recourse to history. As I have suggested, we have numerous historical cases that can well be described as processes of conceptual sedimentation. But it is exactly this material that requires important qualifications. The picture that emerges from these cases is not the formation of a massive bedrock like sandstone, with no possible further movement, in which all gaps are filled. The emerging picture reminds one rather of a porous formation of chalk, or even better a coral reef. On such a reef, we can to a certain degree differentiate between those parts that live and develop and those that have been solidified. But the border between them is not always clear and sharp. Even those coral formations that have sedimented are not absolute and fixed. They may collapse under their own weight, cavities may not only later be filled in, but parts may break off, new vistas open, fragments may attach themselves at other locations, and so on. At many single points on the surface, at different points of time, they offer a base for further attachment; they stimulate specific directions of growth and disfavor other ones. While they offer the base for local development, their development is much too complex to be really determined by its state at a given time. While such a formation indeed develops by a process of sedimentation, it is neither absolute nor immutable. Smaller and major readjustments, even Kuhnian revolutions, are quite possible.

With such a picture I would deviate from what Husserl had in mind with sedimentation and when he talked of a Stufenbau of geometry and the deductive sciences. But Husserl is a complex writer, and still I'm not sure what a Husserl specialist might draw as "his" picture. Such a picture of a coral reef, however, certainly fits well with what Fleck describes as a *Wissensgebäude*, a building that has been anticipated and designed by none of its builders, and that has often been altered in unexpected ways, sometimes indeed against the clear intentions of some of its construction workers (Fleck, 1980, 91). The shift in pictures might well be related to my shift of domains, but one may doubt even this: Modern historiography of mathematics, for example, could probably well sustain such a picture even in the case of the development of mathematical research fields (Epple, 1999).

For our activity as historians and/or philosophers of science, such a picture suggests a further aspect. When we want to explain and understand the structure of any specific reef, we have to apply a peculiar type of explanation. On the one hand, we cannot do without knowing the general laws of crystal growth; the crystallization modes of the materials involved; their mechanical, chemical, thermal, and optical properties; and so on. All this general knowledge is indispensable. At the same time, however, we cannot but have recourse to the reef's specific history, to its specific location, and to chance events: For example, some branches may have started to grow when a ship ran aground on it, destroying some of its parts, and occasioning growth in new directions. A warm ocean current may have changed direction, thus altering the fauna and flora and the process of sedimentation. General laws and the specific history are inseparably intertwined, and we shall never understand the wonderful structure of the reef without invoking such mixed explanations. Both the gross shape of the reef and its individual elements bear traces of its history and cannot be understood without it.

Analogously, we might say that we cannot understand specific scientific concepts without using a mixed method, invoking both general aspects such as appropriateness to explanatory needs and specific historical features of the process of concept generation. This is exactly what Husserl had insisted on toward the end of his *Ursprung* paper: We cannot understand the full meaning of geometrical concepts without looking to their history. To be sure, there are important differences in where exactly this recourse to history might lead us—Husserl's idea of a historical a priori might no longer be shared these days, and chance might remain as an irreducible element. But the common ground is the insight that an understanding of the conceptual structure of science needs a serious look at its history. As Ian Hacking has put it, "concepts have memories" (Hacking, 2002, 37). *And so do facts*, we might now add, with a look at empirical sciences. Both of these statements are at this point probably more like a program of research whose general features and implications remain to be uncovered. The challenge for the future that opens up here might perhaps not be too different from the challenge Husserl wanted to pose to his contemporaries seventy years ago.

Reference Matter

Notes

Introduction

1. *Husserliana, Briefwechsel IX*, pp. 128–29, tr. Engelen.

2. Compare this to Friedrich Steinle's suggestion that it is far *more* common in the experimental sciences than Husserl seems to think.

3. On Foucault, see David Hyder's contribution to this volume.

4. See Rheinberger's and Gasché's contributions to this volume.

Chapter 1

1. See (Smith, 2006) and (2002b) on the unity of theory in Husserl's *Logical Investigations*.

2. See (Smith, 2006) on Husserl's theory of constitution and its background in the mathematical notion of a manifold and early mathematical logic, all relevant to Carnap's program.

3. See (Smith, 2006) and (1995) on this perspectivist reading of Husserl's transcendental idealism; see (Smith, 2006) and (Smith and McIntyre, 1982) on the reconstruction of Husserl's theory of intentionality in a realist, perspectivist scheme aligned with logical-semantic theory.

4. (Carnap, 1928, § 5) following Friedman's translation in (Friedman, 1999, 134), except for putting "constitution theory" for "constitutional theory."

5. (Carnap, 1928, §§ 15–16, 66). See also (Friedman, 1999, 130) as well as (1999, 176).

Chapter 2

Reprinted with permission of *Acta Philosophica Fennica*, originally published in 1990, in *Language, Knowledge, and Intentionality—Perspectives on the*

Philosophy of Jaakko Hintikka, *Acta Philosophica Fennica* 49:123–43. This article springs from a project on Husserl's phenomenology on which I worked as a fellow of the Wissenschaftskolleg, Berlin, in 1989–90. I gratefully acknowledge this support.

1. Simmel conceives of the Lebenswelt as "*aufgebaut*" (constructed) and also writes that "das religiöse Leben *schafft* die Welt" (religious life *creates* the world) (1912, 12, my emphases). Hugo von Hofmannsthal used the term even earlier, in his introduction to the Insel edition of *Thousand and One Nights* (1908), where he wrote: "What would these poems be, what would they mean to us, if they did not emerge from a life-world. This life-world is incomparable, suffused with an infinite joyousness, . . . that brings and binds everything together . . ." (Was wären diese Gedichte, was wären sie uns, wenn sie nicht aus einer Lebenswelt hervorstiegen. Unvergleichlich ist diese Lebenswelt, und durchsetzt von einer unendlichen Heiterkeit, . . . die alles durcheinanderschlingt, alles zueinanderbringt . . .) (von Hofmannsthal 1951, 319, tr. Hyder).

2. See (Merleau-Ponty, 1945, 108n), where he refers to the second volume of the *Ideas*. In the Preface (1945, vii), Merleau-Ponty mentions that he also has studied the manuscript of Husserl's *Crisis*. See (van Breda, 1962), as well as David Carr's English translation of the *Crisis*, p. xxx, notes 20 and 21, and (Sommer, 1990, 84n70).

3. See (Føllesdal, 1982, 553–69).

4. Letter quoted in (Kern, 1964, 276n).

5. Husserl, Preface to the Gibson translation of *Ideas I*. Here from (*Husserliana V*, 152.32–153.5), my translation.

6. This example came up in conversations with David Wellbery.

7. The manuscript dates from 1917 but was copied during the first half of the 1920s, and it is possible that the word Lebenswelt came in then.

8. I have changed Kersten's translation slightly. See (*Husserliana III*, 158.13–19) for the original.

9. Again, the translation was slightly modified. See (*Husserliana III*, 161.15–18) for the original.

10. See, for example, (*Crisis* § 34e, § 36), (*Husserliana VI*, 134, 143, 462).

11. Carr's translation, slightly amended. See (*Husserliana VI*, 143.29–30) for the original.

Chapter 3

1. These lectures will be printed shortly in (Hilbert 2009).
2. See my joint paper with T. Sauer (2006, 213–33).

Chapter 4

1. Husserl in slippers.

2. For references to these and other uses of the word *style* in this tradition, see the opening pages of my "'Style' for Historians and Philosophers" (Hacking, 1992, 1–20). My own theoretical use of the concept "style of scientific reasoning" is adapted from A. C. Crombie on "styles of scientific thinking," much cited in my article. Geoffrey Lloyd, as quoted in the Primal Beginning section above, presumably wrote of "styles of mathematical reasoning" well aware of that discussion.

3. See his contribution to this volume.

4. (Wertheim, 2004).

5. The classic account is found in (Courant and Robbins, 1941) and many later editions.

6. I take the idea of a physicist's tool kit from (Krieger, 1992). When I went to Google to check this reference, I typed in the words "physicist tool kit." My first hit was The Official String Theory Web Site, which begins by answering the question "What is theoretical physics?" with an account of the calculus of variations, Euler and Lagrange.

7. Cf. note 2 above.

Chapter 5

1. See (*The Origin of Geometry*, p. 370).

Chapter 7

First published in Rodolphe Gasché, 2009, *Europe, or The Infinite Task: A Study of a Philosophical Concept* (Stanford, CA: Stanford University Press).

1. This Husserlian caveat is important in many respects: It suggests, in particular, that the very attempt to found Europe upon the Greek idea of philosophy as universal science, and in terms of a universal community, cannot take place in a merely historical fashion. To invoke the tradition in this context, without questioning it critically, is to go against what the very idea of philosophy requires.

2. The naïveté of Greek philosophy is that of its objectivism, but, as Husserl points out, it is of a different kind than that of the modern sciences. See ("Ausdruck. Welt Sinn auflegen," *Husserliana XXIX*, pp. 161–62).

3. "Beilage XXII. Ontologie der Lebenswelt, Ontologie der Menschen," *Husserliana VI*, p. 483, my translation. See also, "Beilage XV. Das europäische

Menschentum in der Krisis der europäischen Kultur. I," *Husserliana VI*, p. 455.

4. For further distinction between imaginary ideality of the morphological type in the pre-geometrical life world and the ideality of pure geometry, see also (Derrida, 1978, 122–26, 133).

5. Husserl makes the distinction when he writes: "What arises first is the idea of continuation which is repeatable with unconditional generality, with its own self-evidence, as a freely thinkable and self-evident possible infinity, rather than the open endlessness [of "imperfect but perfectible subjective representations" of, for example, an individual thing]: rather than finite iteration, this is iteration within the sphere of the unconditional 'again-and-again,' of what can be renewed with ideal freedom" (*Crisis*, *Appendix V*, p. 346).

6. For the connection between idealization, objectification, and method, see (*Crisis*, *Appendix V*, p. 348).

7. In the appendix to the *Crisis* on *The Origin of Geometry*, it is made clear that the very formal-logical self-evidence of all the geometrical propositions that Galileo inherited relieved him from the need to reactivate the actual, that is, the truth-meaning, of geometry (*Origin of Geometry*, pp. 366–67).

8. For a discussion of the notion of the abstract in Husserl, see (Kuhn, 1968, 117).

9. Indeed in *The Vienna Lecture*, Husserl asserts that in the mathematics of antiquity "was accomplished the first discovery of both infinite ideals and infinite tasks. This becomes for all later times the guiding star of the sciences" (*The Vienna Lecture*, p. 293). For a discussion of this apparent contradiction, see (Derrida, 1978, 127). He writes: "Despite the closedness of the system, we are *within* mathematical infinity because we have definitively idealized and gone beyond the factual and sensible finitudes. The infinite infinity of the modern revolution can then be announced in the finite infinity of Antiquity's creation" (130).

10. Is it necessary to point out that Husserl's account, in the *Crisis*, of the birth of the new sciences, in which, for example, the discussion of Galileo dominates at the expense of equally important, and perhaps from a historical perspective, more important figures, is not intended as a history of the sciences or a mundane history of ideas? Indeed, historians of the sciences may object to this privilege accorded to Galileo. Furthermore, they may argue that Galileo's real accomplishment does not consist in the mathematization of nature but in the discovery of the relativity of movement and rest. But the history developed in Part II of the *Crisis* is a history executed in a transcendental-phenomenological attitude, as Elisabeth Ströker notes, that is, "a limine history within the frame of the already presupposed phenomenological epoche" (Ströker, 1979, 113). Husserl's inquiry into the ways Galilean science, which he construes as the

completion of the efforts of various thinkers, such as Vieta, who preceded him, and in which the name *Galileo* does not primarily refer to the historical figure but serves as a designation for an epochal state of mind, is one into "the way of thinking which motivates the idea of the new physics" (*Crisis*, § 9b), its presupposed self-evidences, and "undetermined general anticipations a priori" (*Crisis*, § 9c)—in short, into the subjective processes of this creation. Husserl writes: "Our concern is to achieve complete clarity on the idea and task of a physics which in its Galilean form originally determined modern philosophy, [to understand it] as it appeared in Galileo's own motivation, and to understand what flowed into this motivation from what was traditionally taken for granted and thus remained an unclarified presupposition of meaning, as well as what was later added as seemingly obvious, but which changed its actual meaning" (*Crisis*, § 9e). Thus Husserl can conclude that "in this connection it is not necessary to go more concretely into the first beginnings of the enactment of Galileo's physics and of the development of its method" (ibid.). Inquiring into the intentional structures and the original evidences that constituted the new scientific spirit, the mathematization of nature appears as the very precondition on the basis of which experimentation, as well as the discovery of the relativity of movement and rest, acquire their scientific significance in the first place. To claim that Husserl misrepresents Galileo's achievement misses out on the thrust of Husserl's analysis, the philosophically innovative aspects of his account of the fundamental evidences constitutive of the modern sciences.

11. The point has been made in "Self-Responsibility, Apodicticity, Universality" (Gasché, 2004), and which is included, together with the present essay, in (Gasché, 2009), on the idea of Europe within the phenomenological tradition of philosophy.

12. Holding that the sciences and their history are grounded in the life-world does not amount to relativizing them "in the sense of a social or just epistemological constructivism." As Hans-Jörg Rheinberger has argued, Husserl's aim in grounding the sciences in the life-world is the establishment of a "historical epistemology" (as opposed to pure history of science), an aim that shows Husserl in the proximity of the work of Ludwik Fleck. See (Rheinberger, 2007, 19–21) as well as his contribution to this volume.

13. Husserl writes: "the perfection-limit of the secondary qualities is not measurable; it is only 'intuitable.' But it is intersubjectively determined and determinable through relation to the mathematical limits of the primary characteristics" (*Crisis, Appendix II,* p. 310).

14. Paul Ricoeur also remarks that in spite of its genius, Galileo's "working hypothesis, for lack of self-criticism, is not recognized [by Galileo] as the audacity of spirit at work. Soon this 'indirect mathematization of nature' could verify itself only by the success of its extension, without which the circle of

hypothetical anticipation and unending verification could never be broken, for every enigma of induction is inscribed within this circle" (Ricoeur, 1967, 163).

15. Let us also point out that if the natural sciences achieve compelling apodicticity regarding the idealized spatio-temporal shapes by grounding them logically, it is to be assumed that in the case of idealities that are no longer of the order of bodily shapes, their universal reconstructability may have to have recourse to other than logical means.

16. *Husserliana XXIX*, p. 178, my translation.

Chapter 8

For several valuable comments I would like to thank an anonymous referee and Christian Bermes.

1. Seeking for an origin also has a religious dimension, because it means returning to the point where everything started, to a source of truth, to a point when nothing had been distorted or falsified. David Carr also points this out in his contribution to this volume, when he mentions that phenomenology is related to a return to innocence, to an unfalsified experience.

2. I am grateful to Kevin Mulligan for pointing this out.

3. All of the philosophically relevant origins finally serve as a basis for critical thinking.

4. "Und in unserem Alter einen möglichen Modus des Daseins zu erfinden, nachdem der ganze Boden einem unter den Füßen weggezogen ist, ist eine harte Sache. Das verbraucht viel seelische Kraft, und um damit fertig werden zu können, muß ich mit einer gewaltigen Überkraft philosophischer Konzentration dagegen aufkommen können—und darum dieser äußerste Kampf um die Gestaltung des letzten Werkes" (*Husserliana, Briefwechsel, IX*, pp. 128–29; translated by Hyder).

5. "Ich muß doch schließlich, wenigstens vor mir selbst, rechtfertigen können, daß ich in der deutschen Philosophie (also auch in dieser Nation) kein Fremdling bin, und daß alle die Größen der Vergangenheit, die ich so sehr verehrt habe und deren Gedanken in den meinen in neuen Gestalten wuchsen, mich unbedingt mitrechnen mußten, als echten Erben ihres Geistes, als Blut von ihrem Blute" (*Husserliana, Briefwechsel, IX*, pp. 128–29; translated by Engelen).

6. See for example (Foucault, 1990).

7. Compare for this point the chapters of David Carr and Ulrich Majer in this volume.

8. Ulrich Majer mentions this for Greek mathematics as well in his chapter in this collection.

9. See (*Crisis*, § 9d, 9h).

10. This is quite different with Homer, for whom experience is a bodily process.

11. "How is it that precisely through the epoché a primal ground of immediate and apodictic self-evidences should be exhibited? The answer is: If I refrain from taking any position on the being or non-being of the world, not *every* ontic validity is prohibited for me within this epoché. I, the ego carrying out the epoché, am not included in this realm of objects but rather . . . am excluded in principle. I am necessary as the one carrying it out. It is precisely herein that I find just the apodictic ground I was seeking, the one that absolutely excludes every possible doubt" (*Crisis*, § 17).

12. "And yet, as soon as one took into account that (the exact sciences) are the accomplishments of the consciousness of knowing subjects, their self-evidence and clarity were transformed into incomprehensible absurdity. No offence was taken if, in Descartes, immanent sensibility engendered pictures of the world; but in Berkeley this sensibility engendered the world of bodies itself" (*Crisis*, § 24).

13. "The life-world can be disclosed as a realm of subjective phenomena which have remained 'anonymous'" (*Crisis*, § 29).

14. hoi d' *hama** pantes eph' hippoiin mastigas aeiran,
peplêgon th' *himasin**, homoklêsan t' epeessin
essumenôs: hoi d' *ôka** diepressôn pedioio
(365) nosphi neôn *tacheôs**: hupo de sternoisi koniê
histat' aeiromenê hôs te nephos êe thuella,
chaitai d' *errôonto* meta pnoiêis anemoio+*.
harmata d' allote men chthoni pilnato pouluboteirêi,
allote d' aïxaske metêora: toi d' elatêres
(370) hestasan en diphroisi, patasse de thumos hekastou
nikês hiemenôn: keklonto de hoisin hekastos
hippois, hoi d' epetonto koniontes pedioio.

Then they all at one moment lifted the lash each above his yoke of horses, and smote them with the reins, and called to them with words, full eagerly and forthwith they sped swiftly over the plain [365] away from the ships and beneath their breasts the dust arose and stood, as it were a cloud or a whirlwind, and their manes streamed on the blasts of the wind. And the chariots would now course over the bounteous earth, and now again would bound on high; and they that drove [370] stood in the cars, and each man's heart was athrob as they strove for victory; and they called every man to his horses, that flew in the dust over the plain (Homer, 1925, 23.362–71; translated by Samuel Butler).

Chapter 9

1. Quoted in (Bremer, 1997).
2. For a description of this tension, cf. (Acherman, 2002).

Chapter 10

I wish to thank David Hyder, Skúli Sigurdsson, and an anonymous reviewer for valuable comments and suggestions.

1. It should be mentioned that in this point, Bachelard's attitude is far more complicated.

2. For an early and lucid evaluation of this paper, see Jacob Klein, "Phenomenology and the History of Science" (1940).

3. See David Hyder's contribution to this volume.

4. See Derrida's lengthy introduction to his translation of "The Origin of Geometry," *L'Origine de la géométrie*, itself translated as (Derrida, 1978).

Chapter 11

Originally published in 2003 in *Perspectives on Science* 11 (1):107–29. My thanks to Ian Hacking, Alan Paskow, Alan Richardson, Holger Sturm, and Dieter Teichert for comments on earlier versions of this paper.

1. See Norris's criticisms of Rorty's interpretation of Foucault in (1994, 162–66).

2. One need only think of Wittgenstein's repeated criticisms of *Principia Mathematica*, according to which the inadequacy of Whitehead's and Russell's theory of logic is made evident by their repeated introduction of supplementary, supposedly logical axioms. Wittgenstein's definition of a "tautology" is supposed to provide an independent criterion for distinguishing properly logical propositions; however, his definition presupposes a closed set of bivalent elementary propositions.

3. See Cavaillès's 1938 monograph, *Méthode axiomatique et formalisme* (1981, 165–69).

4. See, for instance, (Gutting, 1989, 9–12.) Gutting takes this passage as his point of departure for a discussion of the theories and methodologies of Bachelard and Canguilhem. Without in any way calling into question the value of his approach (for he rightly emphasizes Foucault's debts within the French tradition of the history of scientific concepts), I am obviously more skeptical of the accu-

racy of Foucault's description. Put otherwise, while Gutting follows Foucault in viewing this French tradition as an *alternative* response to Kant's question "What is Enlightenment?" I am claiming in this essay that Foucault was not able to engage in this alternative critique without covertly borrowing from the transcendental larder. Quite evidently, Foucault's relation to his teachers is far more important when one considers the specific "archaeologies" that he develops.

5. Reprinted as (Foucault, 1994b, 764–65). This article is a redaction of the original version of the preface Foucault had provided for the English translation of Canguilhem's *The Normal and the Pathological.*

6. "Husserl always regretted that an expression that truly captured the aim of philosophy was already in the possession of a positive science, namely the expression: *archaeology*" (Fink, 1939, 246).

7. Who is to be called a structuralist and who not is of course a heated question at this time. Let me just use the term loosely to refer to Foucault, Lacan, Lévi-Strauss, Jakobson, and others at the time who drew on Saussure's work in linguistics.

8. The gap between a subject's immediate grasp of language and the unknown history of that language that Merleau-Ponty takes as the departure for his discussion in (Merleau-Ponty, 1960) is first raised by Hendrik Pos in an article appearing in the *Revue internationale de philosophie* (1939). This issue is dedicated to Husserl, and it contains the first printing of Husserl's *Origin of Geometry*, as well as (Fink, 1939), the article Cavaillès refers to when he describes phenomenology as a kind of archaeology. The concluding paragraphs of Pos's article, in which he stresses the fundamental role of self-consciousness (that is to say, of immediate subjective knowledge) for the *possibility* of the human sciences (his example is linguistics), are an interesting foil to Foucault's later insistence that the human sciences depend on the historically contingent notion of "Man."

9. See my discussion of Frege and Wittgenstein above, as well the passages on logicism in (Cavaillès, 1981, 165–66).

10. (Cavaillès, 1960, 78). Au moins l'évidence justificatrice de l'analyse transcendantale est-elle nécessairement unique: s'il y a conscience des progrès, il n'y a pas progrès de la conscience. Or l'un des problèmes essentiels de la doctrine de la science est que justement le progrès ne soit pas augmentation par juxtaposition, l'antérieur subsistant avec le nouveau, mais révision perpétuelle des contenus par approfondissement et rature.

11. (Cavaillès, 1960, 78). Il n'y a pas une conscience génératrice de ses produits, ou simplement immanente à elles, mais elle est chaque fois dans l'immédiat de l'idée . . . . Le progrès est matériel ou entre essences singulières, son moteur l'exigence de dépassement de chacune d'elles. Ce n'est pas une philosophie de la conscience mais une philosophie du concept qui peut donner une doctrine de la

science. La nécessité génératrice n'est pas celle d'une activité, mais d'une dialectique.

12. Some readers may feel that this scheme of Husserl's is top-heavy. It should be said in his defense, however, that any rational reconstruction implicitly involves the notion of an internal history of concepts that is both distinct of empirical history all while it reveals the true meaning of the latter. That Husserl and Lakatos are obviously inspired by Hegel in this regard may give analytic philosophers reason to reject both of them. But analytic philosophers are just as committed to the notion of "rational history" as was Hegel, and indeed Russell's influential history of Western philosophy was also directly influenced (some might use a stronger term) by Hegel's. Conversely, one cannot very well hope to eliminate such rational historical elements from a history of knowledge in order to do "non-presentist" history of science. As always, Husserl is trying to identify the presuppositions that are involved in a rational investigation of this sort, that is to say, our implicit stance toward past reason.

13. "Dans une philosophie de la conscience ou la logique est transcendantale, ou elle n'est pas" (1960, 26).

14. Some readers may object that Kant has no intention of justifying the principles of logic, but that for him the latter are beyond justification precisely because they are transcendental. It is true that he does not aim to justify his (i.e., our human) logic in the sense of showing it preferable to some alternative. But the critical philosophy does justify the sciences of logic and mathematics in another sense: It guarantees their future validity for our reasoning concerning the empirical world. One can of course restrict the a priori to its constitutive role, thereby "relativizing" it in Reichenbach's sense. But the resulting philosophy, while Kantian, is no longer the philosophy of Kant.

15. (Cavaillès, 1981, 90–93). See also the detailed commentary of (Sinaceur, 1994, 55–66).

16. Cf. (Cavaillès, 1936), quoted and discussed in (Sinaceur, 1994, 108–110).

17. "Si, par la suite, Foucault n'utilise plus le concept d'*a priori historique*, peut-être encore trop évocateur de la phénoménologie, c'est sans doute pour avoir pris conscience non de l'inutilité de l'entreprise, mais de ses limitations. A la fin il l'évoquait, mais de manière programmatique seulement . . . ." (Colette, 1998). I am indebted to Professor Colette for providing me a copy of his unpublished manuscript.

Chapter 12

I would like to thank the Thyssen Foundation (Cologne, Germany) and the Max Planck Institute for the History of Science (Berlin) for generously supporting this research.

1. (Smith and Medin, 1981), (Prinz, 2002), (Gurova, 2003), for example.

2. (Heilbron, 1979) still provides the best overview of electricity of that period.

3. (Heilbron, 1979) chapter 9, gives a brief account of Dufay's research. Dufay's eight "Mémoires" at the Paris Académie between 1733 and 1737 provide the main source; cf. also his own English summary (Dufay, 1734). Electricity here always means, of course, what we nowadays call static electricity.

4. ". . . me déconcerta prodigieusement" (Dufay, 1733b).

5. He mentioned many of them in his "Histoire de l'électricité" (Dufay, 1733a).

6. "die Erfahrung hat uns gelehrt, dass die electrische Kraft von einer zwiefachen Art sey . . ." (Musschenbroek, 1739). Cf. the German translation (Musschenbroek and Gottscheden, 1747, 242).

7. (Winkler, 1745, 5), (Musschenbroek and Gottscheden, 1747, 242), (Gralath, 1747, 208). The latter work is a huge and most comprehensive (natural and chronological) "History of Electricity," which most later historians of electricity, from Priestley to our time, like to use as a source.

8. For the notion of exploratory experimentation, see, e.g., (Steinle, 2003).

9. In recent history of science, by contrast, there is increasing attention to the notion of facts and their historical development, see, e.g. (Shapiro, 2000).

10. My translation.

11. See (*Origin of Geometry*, pp. 360–68).

12. Thanks to the remark of an anonymous referee, I became aware that Husserl sees processes of sedimentation constantly taking place in everyday language as well. This makes the problem of his account of the "descriptive sciences" and of how their language might already be affected by processes of sedimentation even sharper. As far as I see, he does not provide an elaborated account. My interpretation may well go further than the textual evidence can sustain, but its contrasting image provides a background for developing a more differentiated view of Husserl.

13. See Ian Hacking's remarks on "Galileo's mathematization of nature" in his contribution to this volume.

Works by Husserl

Works by Husserl cited in this volume are listed below chronologically. Whenever possible, we use section numbers in preference to page numbers, in order to permit readers to use whichever edition they choose, for example, "(*Crisis*, § 34)." We cite the English translation listed below except when otherwise specified, for instance when referring directly to the Husserliana edition. We use the title or a short form thereof to refer to the edition, for example, "(*Origin of Geometry*, p. 175)" refers to page 175 of Appendix VI to the *Crisis*, "The Origin of Geometry" in David Carr's translation of the *Crisis* listed below. In long or ambiguous cases, we have put the short form used in square brackets, for example, "'Idealization and the Science of Reality—The Mathematization of Nature.' [*Crisis, Appendix II*]" in the list below indicates that this piece is cited in text as "(*Crisis, Appendix II*, p. X)," where again the page number is of the English translation given for that appendix in this list. Since the Appendices of the English translation of the *Crisis* correspond to various Abhandlungen and Beilagen in the Husserliana edition, the numbers of the latter have been given as well, for example, "Appendix VI/Beilage III" refers to the sixth Appendix of the Carr translation and the third Beilage of *Husserliana VI*. When one of these has not been translated, it is listed as "Beilage X" where X denotes its number in *Husserliana VI*. Dual dates, e.g. 1900/2001, give either the original date of publication or, for posthumous works and fragments, the dates of composition, followed by the date of publication of the edition in question. Thus "1891/2003. *Philosophy of Arithmetic*" means that this text was originally published in 1891 and the edition listed is from 2003. "1912–31/1989. *Ideas . . . Second Book*" refers to manuscripts composed by Husserl between 1912 and 1931, published together in 1989 in the edition listed.

Complete Works

1950–2008. *Husserliana: Edmund Husserl—Gesammelte Werke Vol. I–XL.* Edited by Ullrich Melle. The Hague: Matinus Nijhoff. [*Husserliana*].

Individual Works

1891/1970. *Philosophie der Arithmetik. Husserliana XII.* Edited by Lothar Eley. The Hague: Martinus Nijhoff.

1891/2003. *Philosophy of Arithmetic: Psychological and Logical Investigations with Supplementary Texts from 1887–1901.* Translated by D. Willard. Dordrecht, Netherlands, and Boston: Kluwer Academic Publishers.

1900–1901/2001. *Logical Investigations, Volumes One and Two.* Translated by J. N. Findlay. Edited and Revised by Dermot Moran. London and New York: Routledge.

1910–11/1965. "Philosophy as Rigorous Science." In *Phenomenology and the Crisis of Philosophy.* Translated by Quentin Lauer. New York: Harper.

1912–31/1989. *Ideas pertaining to a Pure Phenomenology and to a Phenomenological Philosophy, Second Book: Studies in the Phenomenology of Constitution.* Translated by Richard Rojcewicz and André Schuwer. Dordrecht, Netherlands, and Boston: Kluwer. [*Ideas II*]

1913/1931. *Ideas pertaining to a Pure Phenomenology and a Phenomenological Philosophy, First Book: General Introduction to Pure Phenomenology.* Translated by W. R. Boyce Gibson. London: George Allen & Unwin Ltd., and New York: Humanities Press. [*Ideas I*]

1928–37/1970. *The Crisis of European Sciences and Transcendental Phenomenology. An Introduction to Phenomenological Philosophy.* Translated by David Carr. Evanston, IL: Northwestern University Press. [*Crisis*]

1928–37/1976. *Die Krisis der europäischen Wissenschaften und die transzendentale Phänomenologie. Eine Einleitung in die phänomenologische Philosophie. Husserliana VI.* Edited by Walter Biemel. The Hague: Martinus Nijhoff. [*Husserliana VI*]

1929/1969. *Formal and Transcendental Logic.* Translated by Dorion Cairns. The Hague: Martinus Nijhoff.

1931/1960. *Cartesian Meditations.* Translated by Dorion Cairns. Dordrecht: Kluwer.

1936. "Die Krisis der europäischen Wissenschaften und die transzendentale Phänomenologie. Eine Einleitung in die phänomenologische Philosophie." *Philosophia I* (1936): 77–176.

1934–37/1993. *Die Krisis der europäischen Wissenschaften und die transzendentale Phänomenologie. Ergänzungsband. Texte aus dem Nachlass. 1934–1937.*

Husserliana XXIX. Edited by Reinhold N. Smid. Dordrecht, Netherlands: Kluwer.

APPENDICES TO THE *CRISIS*

Appendix I/Abhandlung III

1935/1970. "The Vienna Lecture." In *The Crisis of European Sciences and Transcendental Phenomenology*. Translated by David Carr. Evanston, IL: Northwestern University Press. [*The Vienna Lecture*]

Appendix II/Abhandlung I

1928/1970. "Idealization and the Science of Reality—The Mathematization of Nature." In *The Crisis of European Sciences and Transcendental Phenomenology*. Translated by David Carr. Evanston, IL: Northwestern University Press. [*Crisis, Appendix II*]

Appendix V/Beilage II

1936/1970. "Objectivity and the World of Experience." In *The Crisis of European Sciences and Transcendental Phenomenology*. Translated by David Carr. Evanston, IL: Northwestern University Press. [*Crisis, Appendix V*]

Appendix VI/Beilage III

1939. "Die Frage nach dem Ursprung der Geometrie als intentionalhistorisches Problem." *Revue Internationale de Philosophie* 1 (2):203–25.

1939/1962. *L'Origine de la géométrie*. Translated by J. Derrida. Paris: Presses Universitaires de France.

1939/1970. "The Origin of Geometry." In *The Crisis of European Sciences and Transcendental Phenomenology*. Translated by David Carr. Evanston, IL: Northwestern University Press. [*Origin of Geometry*]

Appendix VII/Beilage XVII

1936–37/1970. "The Life World and the World of Science." In *The Crisis of European Sciences and Transcendental Phenomenology*. Translated by David Carr. Evanston, IL: Northwestern University Press. [*Crisis, Appendix VII*]

Beilage XV

1936–37/1976. "Das europäische Menschentum in der Krisis der europäischen Kultur. I." In *Die Krisis der europäischen Wissenschaften und die transzendentale Phänomenologie. Husserliana VI*. Edited by Walter Biemel. The Hague: Martinus Nijhoff.

Beilage XXIII

1936/1976. "Ontologie der Lebenswelt, Ontologie der Menschen." In *Die Krisis der europäischen Wissenschaften und die transzendentale Phänomenologie. Husserliana VI*. Edited by Walter Biemel. The Hague: Martinus Nijhoff.

1938/1973. *Experience and Judgment: Investigations in a Genealogy of Logic.* Translated by James S. Churchill and Karl Ameriks. Evanston, IL: Northwestern University Press.

Correspondence

1994. *Edmund Husserl: Briefwechsel. Husserliana: Edmund Husserl Dokumente. Vol. III/I–X.* Edited by Elisabeth and Karl Schuhmann. Berlin: Springer. [*Husserliana, Briefwechsel*]

Unpublished Work

"Manuscript B II 9." Quoted in (Diemer, 1965, 68n29).

General Bibliography

Achermann, Eric. 2002. Existieren Epochen? *Mitteilungen des Deutschen Germanistenverbandes* 49 (3):222–39.

Arbman, Ernst. 1926. Untersuchungen zur primitiven Seelenvorstellung mit besonderer Rücksicht auf Indien: pt 1/2. *Le Monde Oriental* 20:85–222.

Avenarius, Richard. 1891. *Der menschliche Weltbegriff.* Leipzig, Germany: Reisland.

Bachelard, Gaston. 1969. *La formation de l'esprit scientifique.* Paris: Vrin.

———. 1987. *Essai sur la connaissance approchée.* Paris: Vrin.

Boi, Luciano. 2004. Theories of Space-Time in Modern Physics. *Synthese* 139 (3):429–89.

Boltzmann, Ludwig. 1905. Der zweite Hauptsatz der mechanischen Wärmetheorie. *Populäre Schriften.* Leipzig: Barth.

Bolzano, Bernard. 1972. *Theory of Science: Attempt at a Detailed and in the Main Novel Exposition of Logic with Constant Attention to Earlier Authors.* Translated and edited by R. George. Berkeley: University of California Press.

Bose, Georg Matthias. 1744. *Tentamina electrica in academiis regiis Londinensi et Parisana primum habita, omni studio repetita.* Wittenberg, Germany: Johann Ahlfeld.

Breda, H. L. van. 1962. Maurice Merleau-Ponty et les Archives-Husserl à Louvain. *Revue de métaphysique et de morale* 67:410–30.

Bremer, Józef. 1997. *Rekategorisierung statt Reduktion. Zu Wilfrid Sellars' Philosophie des Geistes.* Göttingen, Germany: Vandenhoeck & Ruprecht.

Canguilhem, Georges. 1967. Mort de l'homme ou épuisement du cogito. *Critique* 242:599–618.

———. 1977. *Idéologie et rationalité dans l'histoire des sciences de la vie.* Paris: Vrin.

Carnap, Rudolf. 1928. *Der logische Aufbau der Welt.* Berlin: Weltkreis.

———. 2003. *The Logical Structure of the World.* Translated by R. George. La Salle, IL: Open Court.

Carr, David. 1987. *Interpreting Husserl.* Dordrecht, Netherlands: Nijhoff.

Cavaillès, Jean. 1936. Le congrès international de philosophie des sciences. *Revue de métaphysique et de morale* 43:118–20.

———. 1960. *Sur la logique et la théorie des sciences.* Paris: Presses universitaires de France.

———. 1981. *Méthode axiomatique et formalisme.* Paris: Hermann.

Chemla, Karine, and Shuchun Guo, eds. 2004. *Les neufs chapitres: le classique mathématique de la Chine.* Paris: Dunod.

Colette, Jacques. 1998. Paradoxes et apories de la notion d'a priori historique: E. Husserl 1929–36. M. Foucault 1966–69. Unpublished manuscript of lectures delivered in Brussels, December 11, 1997, and Paris, February 6, 1998.

Courant, Richard. 1937. *Differential and Integral Calculus.* Translated by E. J. McShane. Glasgow, UK: Blackie.

Courant, Richard, and Herbert Robbins. 1941. *What is Mathematics?* New York: Oxford University Press.

Dennett, Daniel. 1991. *Consciousness Explained.* Boston: Little, Brown and Company.

Derrida, Jacques. 1978. *Edmund Husserl's Origin of Geometry: An Introduction.* Translated by J. P. Leavey. Stony Brook, NY: Nicholas Hays, Ltd.

———. 1999. *Sur parole. Instantanés philosophiques.* Paris: Editions de l'Aube.

Diemer, Alwin. 1965. *Edmund Husserl. Versuch einer systematischen Darstellung seiner Phänomenologie.* Meisenheim am Glan: Anton Hain.

Dilthey, Wilhelm. 1968. *Der Aufbau der geschichtlichen Welt in den Geisteswissenschaften. Gesammelte Schriften Vol. VII.* Edited by B. Groethuysen. 5th ed. Stuttgart, Germany: B. Teubner.

Doppelmayr, Johann Gabriel. 1744. *Neu-entdeckte Phaenomena von bewunderungswürdigen Würkungen der Natur.* Nürnberg, Germany: Fleischmann.

Du Bois-Reymond, Emil. 1912. Über Geschichte der Wissenschaft. In *Reden: In zwei Bänden Vol. 1.* Edited by E. Du Bois-Reymond. Leipzig, Germany: Veit & Co.

Dufay, Charles François de Cisternai. 1733a. Premier mémoire sur l'électricité. Histoire de l'électricité. *Histoire de l'Académie Royale des Sciences, avec les Mémoires de Mathématique & de Physique pour la même année.*

———. 1733b. Quatrième mémoire sur l'électricité. De l'attraction et répulsion des corps électriques. *Histoire de l'Académie Royale des Sciences, avec les Mémoires de Mathématique & de Physique pour la même année.*

———. 1734. A Letter from Mons. Du Fay, F. R. S. and of the Royal Academy of Sciences at Paris, to His Grace Charles Duke of Richmond and Lenox, concerning Electricity. Translated from the French by T. S. M. D. *Philosophical Transactions* 38 (431):258–66.

Epple, Moritz. 1999. *Die Entstehung der Knotentheorie: Kontexte und Konstruktionen einer modernen mathematischen Theorie.* Braunschweig, Germany: Vieweg.

Fink, Eugen. 1939. Das Problem der Phänomenologie Edmund Husserls. *Revue internationale de philosophie* 1 (2):226–70.

Fleck, Ludwik. 1929. Zur Krise der "Wirklichkeit." *Die Naturwissenschaften* 17:425–30.

———. 1935. *Entstehung und Entwicklung einer wissenschaftlichen Tatsache. Einführung in die Lehre vom Denkstil und Denkkollektiv.* Basel, Switzerland: Schwabe.

———. 1979. *The Genesis and Development of a Scientific Fact.* Chicago: University of Chicago Press.

———. 1980. *Entstehung und Entwicklung einer wissenschaftlichen Tatsache: Einführung in die Lehre vom Denkstil und Denkkollektiv.* Frankfurt am Main, Germany: Suhrkamp.

Fodor, Jerry. 1982. Methodological Solipsism Considered as a Research Strategy in Cognitive Psychology. In *Husserl, Intentionality and Cognitive Science.* Cambridge, MA: MIT Press.

Føllesdal, Dagfinn. 1979. Husserl and Heidegger on the Role of Actions in the Constitution of the World. In *Essays in Honour of Jaakko Hintikka*. Edited by E. Saarinen, R. Hilpinen, I. Niiniluoto, and M. Provence Hintikka. Dordrecht, Netherlands: Reidel.

———. 1982. Intentionality and Behaviorism. In *Proceedings of the 6th International Congress of Logic, Methodology and Philosophy of Science, Hannover, August 22–29, 1979.* Edited by L. J. Cohen, J. Los, H. Pfeiffer, and K.-P. Podewski. Amsterdam: North-Holland.

———. 1988. Husserl on Evidence and Justification. In *Edmund Husserl and the Phenomenological Tradition: Essays in Phenomenology.* Edited by R. Sokolowski. Washington, DC: The Catholic University of America Press.

Foucault, Michel. 1966. *Les mots et les choses.* Paris: Gallimard.

———. 1969. *L'archéologie du savoir.* Paris: Gallimard.

———. 1990. Qu'est ce que la critique? *Bulletin de la Société francaise de Philosophie* 84.

———. 1994a. Foucault répond à Sartre. In *Dits et Écrits Vol. 1.* Edited by D. Defert and F. Ewald. Paris: Gallimard.

———. 1994b. La vie: l'expérience et la science. In *Dits et Écrits Vol. 4.* Edited by D. Defert and F. Ewald. Paris: Gallimard.

Frercks, Jan. 2004. Disziplinenbildung und Vorlesungsalltag: Funktionen von Lehrbüchern der Physik um 1800 mit einem Fokus auf die Universität Jena. *Berichte zur Wissenschaftsgeschichte* 27:27–53.

Friedman, Michael. 1999. *Reconsidering Logical Positivism*. Cambridge and New York: Cambridge University Press.

———. 2000. *A Parting of the Ways: Carnap, Cassirer, and Heidegger*. Chicago and LaSalle, IL: Open Court.

———. 2001. *Dynamics of Reason: The 1999 Kant Lectures at Stanford University*. Stanford, CA: CSLI Publications.

Gasché, Rudolphe. 2004. Self-Responsibility, Apodicticity, Universality. *Menschenontologie* 10:17–37.

———. 2009. *Europe, or the Infinite Task: A Study of a Philosophical Concept*. Stanford, CA: Stanford University Press.

Gralath, Daniel. 1747. Geschichte der Electricität. *Versuche und Abhandlungen der Naturforschenden Gesellschaft zu Danzig* 1:175–304.

Guericke, Otto von. 1672. *Ottonis de Guericke experimenta nova (ut vocantur) Magdeburgica de vacuo spatio, primum a Gaspare Schotto, nunc vero ab ipso auctore perfectius edita, variisque aliis experimentis aucta*. Amstelodami: Joannus Janssonius.

Gurova, Lilia. 2003. Philosophy of Science Meets Cognitive Science: The Categorization Debate. In *Bulgarian Studies in the Philosophy of Science*. Edited by D. Ginev. Dordrecht, Netherlands: Kluwer.

Gutting, Gary. 1989. *Michel Foucault's Archaeology of Scientific Reason*. Cambridge: Cambridge University Press.

Hacking, Ian. 1992. "Style" for Historians and Philosophers. *Studies in History and Philosophy of Science* 23:1–20.

———. 2000. What Mathematics Has Done to Some and Only Some Philosophers. In *Mathematics and Necessity*. Edited by T. J. Smiley. London: British Academy.

———. 2002. *Historical Ontology*. Cambridge, MA: Harvard University Press.

Harrington, Anne. 1996. *Reenchanted Science: Holism in German Culture from Wilhelm II to Hitler*. Princeton, NJ: Princeton University Press.

Hartmann, Max. 1927. *Allgemeine Biologie*. Jena: Fischer.

———. 1956. Die Kausalität in Physik und Biologie. In *Gesammelte Vorträge und Aufsätze: II. Naturphilosophie*. Stuttgart, Germany: Gustav Fischer.

Heidegger, Martin. 1927. *Sein und Zeit*. Tübingen, Germany: Niemeyer.

Heilbron, John L. 1979. *Electricity in the 17th and 18th Centuries*. Berkeley: University of California Press.

Hilbert, David. 1919. Natur und mathematisches Erkennen. Unedited manuscript preserved at the Mathematisches Institut der Universität Göttingen.

———. 1924. Über die Einheit in der Naturerkenntnis. Unedited manuscript preserved at the Mathematisches Institut der Universität Göttingen.

———. 2009. *Lectures on the Foundations of Physics: Writings after 1914*. Berlin: Springer.

Hofmannsthal, Hugo von. 1951. *Prosa II.* Frankfurt am Main, Germany: Fischer.
Homer. 1925. *The Iliad.* Translated by S. Butler. London: Jonathan Cape.
Kant, Immanuel. 1924. *Critique of Pure Reason.* Translated by J. M. D. Meiklejohn. London: G. Bell & Sons Ltd.
Kern, Iso. 1964. *Husserl und Kant. Eine Untersuchung über Husserls Verhältnis zu Kant und zum Neukantianismus, (Phenomenologica 16).* The Hague: Martinus Nijhoff.
Klein, Jacob. 1940. Phenomenology and the History of Science. In *Philosophical Essays in Memory of Edmund Husserl.* Edited by M. Farber. Cambridge, MA: Harvard University Press.
Klein, Ursula. 1994. Origin of the Concept of Chemical Compound. *Science in Context* 7 (2):163–204.
Krieger, Martin H. 1992. *Doing Physics: How Physicists Take Hold of the World.* Bloomington: Indiana University Press.
Kuhn, Helmut. 1968. The Concept of Horizon. In *Philosophical Essays in Memory of Edmund Husserl.* Edited by M. Farber. New York: Greenwood Press.
Latour, Bruno. 1990. The Force and the Reason of Experiment. In *Experimental Inquiries: Historical, Philosophical and Social Studies of Experimentation in Science.* Edited by H. E. Le Grand. Dordrecht, Netherlands: Kluwer.
Liebert, Arthur. 1923. *Die Geistige Krisis der Gegenwart.* Berlin: Pan-Verlag.
Luft, Sebastian. 2004. Die Archivierung des husserlschen Nachlasses 1933–1935. *Husserl Studies* 20:1–23.
Majer, Ulrich, and Tilman Sauer. 2006. Intuition and Axiomatic Method in Hilbert's Foundation of Physics: Hilbert's Idea of a Recursive Epistemology in his Third Hamburg Lecture. In *Intuition and the Axiomatic Method.* Edited by E. Carson and R. Huber. Dordrecht, Netherlands: Springer Netherlands.
Merleau-Ponty, Maurice. 1945. *Phénoménologie de la perception.* Paris: Gallimard.
———. 1960. Sur la phénoménologie du langage. In *Signes.* Paris: Gallimard.
Musschenbroek, Pieter (Petrus) van. 1739. *Essai de physique—Beginsels der natuurkunde.* Leyden, Netherlands: Luchtmans.
Musschenbroek, Pieter (Petrus) van, and Johann Christoph Gottscheden. 1747. *Grundlehren der Naturwissenschaft.* Leipzig, Germany: Kiesewetter.
Netz, Reviel. 1999. *The Shaping of Deduction in Greek Mathematics: A Study in Cognitive History.* Cambridge: Cambridge University Press.
Norris, Christopher. 1994. "What Is Enlightenment?" Kant and Foucault. In *The Cambridge Companion to Foucault.* Edited by G. Gutting. Cambridge: Cambridge University Press.

Overbye, Dennis. 2004. From Companion's Lost Diary, A Portrait of Einstein in Old Age. *New York Times*, April 24, 2004.

Pannwitz, Rudolf. 1917. *Die Krisis der Europäischen Kultur, Werke Vol. II.* Nuremberg, Germany: H. Carl.

Patočka, Jan. 1988. La philosophie de la crise des sciences d'après Edmund Husserl et sa conception d'une phénoménologie du "monde de la vie." In *Le monde naturel et le mouvement de l'existence humaine.* Dordrecht, Netherlands: Kluwer.

Pos, H.-J. 1939. Phénoménologie et linguistique. *Revue internationale de philosophie* 1 (2):354–65.

Prinz, Jesse J. 2002. *Furnishing the Mind: Concepts and Their Perceptual Basis, Representation and Mind.* Cambridge, MA: MIT Press.

Quine, Willard Van Orman. 1980. Two Dogmas of Empiricism. In *From a Logical Point of View.* Cambridge, MA: Harvard University Press.

Quine, Willard Van Orman, and J. S. Ullian. 1978. *The Web of Belief.* 2nd ed. New York: Random House.

Rappe, Guido. 1996. Das Herz im Kulturvergleich. In *Das Herz im Kulturvergleich.* Edited by G. Berkemer and G. Rappe. Berlin: Akademie Verlag.

Rheinberger, Hans-Jörg. 2005. Gaston Bachelard and the Notion of "Phenomenotechnique." *Perspectives on Science* 13 (3):313–28.

———. 2007. Zur Historizität wissenschaftlichen Wissens: Ludwik Fleck, Edmund Husserl. In *Krise des Historismus— Krise der Wirklichkeit, Wissenschaft, Kunst und Literatur.* Edited by O. G. Oexle. Göttingen, Germany: Vandenhoeck & Ruprecht.

Ricoeur, Paul. 1967. *Husserl: An Analysis of His Phenomenology.* Translated by E. G. Ballard and L. E. Embree. Evanston, IL: Northwestern University Press.

Riezler, Kurt. 1928. Die Krise der "Wirklichkeit." *Die Naturwissenschaften* 16:705–12.

Schiemann, Gregor. 1997. Geschichte und Natur zwischen Differenz und Konvergenz. In *Geschichtsdiskurs Bd. 4: Krisenbewußtsein, Katastrophenerfahrungen und Innovationen 1880-1945.* Edited by W. Küttler, J. Rüsen, and E. Schulin. Frankfurt am Main, Germany: Fischer.

Schlesier, Renate. 2000. Die dionysische Psyche. Zu Euripides' Bakchen. In *Emotionalität. Zur Geschichte der Gefühle.* Edited by C. Benthien, A. Fleig, and I. Kasten. Köln, Germany: Böhlau Verlag.

Sellars, Wilfrid. 1962. Philosophy and the Scientific Image of Man. In *Frontiers of Science and Philosophy.* Edited by R. Colodny. Pittsburgh, PA: University of Pittsburgh Press.

———. 1963. *Science, Perception and Reality.* London: Routledge and Kegan Paul.

———. 1964. Introduction to the Philosophy of Science. Lectures given at the Summer Institute for the History of Philosophy of Science at the American University, Washington, DC, June 1964. Typescript (40 pages).

———. 1981. The Carus Lectures. Foundations for a Metaphysics of Pure Process. *The Monist* 64:3–90.

Shapiro, Barbara J. 2000. *A Culture of Fact: England, 1550–1720*. Ithaca, NY: Cornell University Press.

Simmel, Georg. 1912. *Die Religion*. Edited by M. Buber. 2nd ed. Frankfurt am Main, Germany: Rütten & Loening.

Sinaceur, Hourya. 1994. *Jean Cavaillès, philosophie, mathématique*. Paris: Presses Universitaires de France.

Smith, David Woodruff. 1989. *The Circle of Acquaintance: Perception, Consciousness, and Empathy*. Dordrecht, Netherlands and Boston: Kluwer.

———. 1995. Mind and Body. In *The Cambridge Companion to Husserl*. Edited by B. Smith and D. W. Smith. Cambridge and New York: Cambridge University Press.

———. 2002a. Mathematical Form in the World. *Philosophia Mathematica* 10 (3):102–29.

———. 2002b. What is 'Logical' in Husserl's *Logical Investigations*? The Copenhagen Interpretation. In *One Hundred Years of Phenomenology: Husserl's Logical Investigations Revisited*. Edited by D. Zahavi and F. Stjernfelt. Dordrecht, Netherlands and Boston: Kluwer.

———. 2006. *Husserl*. London and New York: Routledge.

Smith, David Woodruff, and Ronald McIntyre. 1982. *Husserl and Intentionality: A Study of Mind, Meaning, and Language*. Boston and Dordrecht, Netherlands: Reidel and Kluwer.

Smith, Edward E., and Douglas L. Medin. 1981. *Categories and Concepts (Cognitive Science Series)*. Cambridge, MA: Harvard University Press.

Sommer, Manfred. 1990. *Lebenswelt und Zeitbewußtsein*. Frankfurt am Main, Germany: Suhrkamp.

Steinle, Friedrich. 1996. Work, Finish, Publish? The Formation of the Second Series of Faraday's "Experimental Researches in Electricity." *Physis* 33:141–220.

———. 2002. Challenging Established Concepts: Ampère and Exploratory Experimentation. *Theoria: revista de teoria, historia y fundamentos de la ciencia* 17:291–316.

———. 2003. Experiments in History and Philosophy of Science. *Perspectives on Science* 10 (4):408–32.

———. 2005. *Explorative Experimente. Ampère, Faraday und die Ursprünge der Elektrodynamik*. Stuttgart, Germany: Franz Steiner Verlag.

Ströker, Elisabeth. 1979. Geschichte und Lebenswelt als Sinnesfundament der Wissenschaften in Husserls Spätwerk. In *Lebenswelt und Wissenschaft in der Philosophie Edmund Husserls*. Edited by E. Ströker. Frankfurt am Main, Germany: Klostermann.

———. 1997. *Husserlian Foundations of Science*. 2nd ed. Berlin and New York: Springer.

Troeltsch, Ernst. 1922a. *Der Historismus und seine Probleme*. Tübingen, Germany: J. C. B. Mohr.

———. 1922b. Die Krisis des Historismus. *Neue Rundschau* 33:572–90.

Wertheim, Margaret. 2004. Cones, Curves, Shells, Towers: He Made Paper Jump to Life. *The New York Times*, June 22, 2004.

Weyl, Hermann. 1968. Wissenschaft als symbolische Konstruktion des Menschen. In *Gesammelte Abhandlungen Vol. 4*. Edited by K. von Chandrasekharan. Berlin: Springer.

Winkler, Johann Heinrich. 1745. *Die Eigenschaften der electrischen Materie und des electrischen Feuers, aus verschiedenen neuen Versuchen erkläret, und, nebst etlichen neuen Maschinen zum Electrisiren beschrieben*. Leipzig, Germany: Breitkopf.

Index